St. Louis Community College

WITHDRAWN

FV

 St. Louis Community College

Forest Park
Florissant Valley
Meramec

Instructional Resources
St. Louis, Missouri

The potato in the human diet

JENNIFER A. WOOLFE
with contributions from Susan V. Poats

Published in collaboration with
INTERNATIONAL POTATO CENTER

CAMBRIDGE UNIVERSITY PRESS
Cambridge
London   New York   New Rochelle
Melbourne   Sydney

Published by the Press Syndicate of the University of Cambridge
The Pitt Building, Trumpington Street, Cambridge CB2 1RP
32 East 57th Street, New York, NY 10022, USA
10 Stamford Road, Oakleigh, Melbourne 3166, Australia

International Potato Center, PO Box 5969, Lima, Peru

© Cambridge University Press 1987

First published 1987

Printed in Great Britain at the University Press, Cambridge

British Library cataloguing in publication data

Woolfe, Jennifer A.
  Potato in the human diet.
  1. Potatoes 2. Nutrition
  I. Title II. Poats, Susan V.
  641.3′521 TX401

Library of Congress cataloguing in publication data

Woolfe, Jennifer A.
  Potato in the human diet.
  Includes bibliographies and index.
  1. Potatoes. I. Poats, Susan V. II. Title.
TX558.P8W63 1987 641.3′521 86-23311

ISBN 0 521 32669 9

Contents

Preface	xi
Acknowledgements	xiii
Abbreviations and terms	xv
Introduction	1
Current and future roles	1
Misconceptions and a remedy	3
An outline of the book	3
References	6
1 Structure of the potato tuber and composition of tuber dry matter	7
Structure of the tuber	7
Dry matter	8
Carbohydrates	10
Nitrogen	12
Lipids	12
Enzymes	13
Organic acids	14
Pigments	14
Vitamins	15
Ash	16
Concluding remarks	16
References	16
2 The nutritional value of the components of the tuber	19
Dry matter	23
Energy and protein	24
Energy value	24
Protein content	30
Dietary fibre	38
Vitamins	40
Factors affecting contents	40
Contribution to the diet	44

Minerals and trace elements	47
Factors affecting contents	48
Contribution to the diet	48
Summary	51
References	54
3 Protein and other nitrogenous constituents of the tuber	**58**
Part 1. Composition of tuber nitrogen	58
Factors affecting total tuber nitrogen	58
Constituents	60
Soluble protein	61
Non-protein nitrogen	62
Amino acid composition of the whole tuber	64
Part 2. Nutritive value of tuber nitrogen	66
Amino acid analyses and scores	66
Microbiological assays and animal feeding experiments	68
Human feeding experiments	71
Adults	71
Children	72
Comments on protein contribution from potatoes	74
Part 3. Potato protein from processing waste	75
Reasons for waste production	75
Protein recovery	76
Nutritional value	76
Use in food for humans	77
References	78
4 Effects of storage, cooking and processing on the nutritive value of potatoes	**83**
Part 1. Storage	85
Storage conditions	85
Changes in potato nutrients as a result of storage	90
Carbohydrates	90
Nitrogenous constituents	90
Fibre	93
Vitamins	93
Minerals	98
Comments on nutritional changes due to storage	99
Part 2. Main methods of domestic preparation	100
Peeling	100
Nutritional aspects	101
Distribution of nutrients in the tuber	101
Boiling unpeeled *versus* peeled potatoes	103
Moisture	104
Carbohydrate	104
Nitrogenous constituents	106

Contents vii

Fibre	107
Vitamins	107
Minerals	110
Other domestic methods of potato preparation	110
Nitrogenous constituents	111
Fibre	112
Vitamins	112
Minerals	116
Comments on nutritional changes during domestic cooking	117
Part 3. Processing	118
Large-scale processing	119
Pre-peeled potatoes	119
Frozen potato products	123
Potato chips (crisps)	128
Dehydrated potato products	131
Canned potatoes	136
Comments on nutritional changes during processing	139
Traditional processing	143
Production of chuño	144
Papa seca	148
Nutritive value of traditionally dried potato products	149
Part 4. Summary	152
References	154
5 Glycoalkaloids, proteinase inhibitors and lectins	162
Glycoalkaloids	162
Chemical structure and content in tuber	162
Physiological functions	169
Effect on potato flavour	170
Accumulation	172
Toxicity	177
Control of accumulation	179
Proteinase inhibitors	181
Chemical structure and functions	181
Nutritional significance	183
Lectins	184
Summary	185
References	186
6 Patterns of potato consumption in the tropics	191
Calculating consumption	192
Food balance sheets	192
Nutrition surveys	194
Consumer surveys and consumer groups	195
Results of consumer surveys in selected countries	199
Indonesia	199

Rwanda	202
Guatemala	203
Peru	204
How and why potatoes are consumed	206
A typology of potato consumption	208
Consumption role and potato price	209
Other factors influencing potato consumption	211
Prospects	214
Potential for changes in consumption roles	214
Processing	215
Weaning foods	217
The place of potatoes in vegetable production	220
Conclusions	220
References	221
Index	222

'Oh, it will not bear polish, the ancient potato
Needn't be nourished by Caesars, will blow anywhere
Hidden by nature, counted-on, stubborn and blind.

You may have noticed the bush it pushes to air,
Comical, delicate, sometimes with second-rate flowers
Awkward and milky and beautiful only to hunger.'

Richard Wilbur, *The beautiful changes and other poems*
(Harcourt, Brace, Jovanovich Inc., 1947)

Preface

The idea for this book arose from the large number of requests to the International Potato Center (Centro Internacional de la Papa) for information on consumption and nutritional aspects of potatoes. There was clearly a need for an up-to-date review, particularly in respect of developing countries. Within its mandate to disseminate information on potatoes, the International Potato Center funded this review of the potato's nutritional value. The work was part of a larger three-year project on potato consumption and utilization in developing countries carried out by Dr Susan Poats.

Over 700 titles concerning various aspects of the potato as a food were collected, and Chapters 1 to 5 survey this literature. Because there are few data available on potato consumption in developing countries, Chapter 6 presents the results of some case studies in the tropics by Dr Poats.

I hope that the book provides useful information for, and stimulus to, the work of all those concerned with the greater exploitation of the potato as a food contributing significantly to the health and well-being of humankind. It may also be of value to the interested casual reader who simply wishes to learn more about the dietary role of potatoes.

January 1986   J.A.W.

Acknowledgements

My thanks are due primarily to the main contributor to this book, Dr Susan Poats. She inspired the idea of the book, helped to plan much of the initial outline, and was responsible for much of Chapter 5 and many of the photographs, which she took whilst employed by the International Potato Center. She also allowed me to use her findings on potato consumption, which were the fruit of several years work, for Chapter 6.

Mrs Carla Fjeld made helpful comments on the style of the manuscript and provided very useful ideas for Chapter 2. Her inspiring sense of enthusiasm and her encouragement during the final part of preparing the review are greatly appreciated.

I am particularly grateful to Professor Arnold Bender and Dr Glynn Burton for reading the manuscript. Their numerous expert comments, corrections and suggestions have enabled me to improve both its style and content.

Thanks are also due to Ms Kerstin Olsson for reading, and providing useful comments on, that part of Chapter 5 dealing with potato glycoalkaloids, and to Dr Robert Booth and Mr Peter Keane for reading, and commenting on, Chapter 4.

I acknowledge the tremendous amount of time and assistance which the library staff of the International Potato Center has given to this project. In particular I thank Ms Carmen de Podestá, Ms Cecilia Ferreyra and Mr Feliciano Orellana for their unfailing patience with my requests for documents. I am also grateful to Ms Paulette George and the Post Harvest Institute for Perishables for their assistance with literature searches and provision of some documents unavailable in Peru.

This work was carried out as part of the activities of the Social Science Department of the International Potato Center. I would like to express my gratitude to the head of department, Dr Douglas Horton, for his continual support, interest and enthusiasm throughout the preparation of this publication, and to the secretarial staff for the time and patience they devoted to typing the manuscript.

Thanks are also due to Mrs Sandi Irvine for editing, and to Dr Lilo Schilde and Dr Peter Schmiediche for translation of German documents into English, to Dr Orville Page and Dr José Valle Riestra for reading and commenting on parts of the manuscript, to Ms Jésus Chang and Mr Abel Mondragón for help with choosing and preparing the photographs, and to Ms Linda Peterson and other friends and colleagues at the International Potato Center who have contributed with suggestions and help.

Abbreviations and terms

AIS-N: alcohol-insoluble nitrogen
DM: dry matter
DWB: dry weight basis
FAO: Food and Agriculture Organization
FBS: food balance sheet
FWB: fresh weight basis
LSG: low specific gravity
HSG: high specific gravity
N: nitrogen
NDpCal%: net dietary protein calories percentage
NDpER: net dietary protein energy ratio ($0.01 \times$ NDpCal%)
NPN: non-protein nitrogen
PPC: potato protein concentrate
RDA: recommended daily allowance
USRDA: United States recommended daily allowance
WHO: World Health Organization

Biological value (BV): the proportion of absorbed nitrogen which is retained in the body for maintenance and/or growth.

Chemical or protein score: the limiting amino acid in a test protein expressed as a percentage of the same amino acid in a standard (egg or a reference protein)

Essential amino acid index (EAA index): the geometric mean of the ratios of the essential amino acids in a protein to those of a standard (usually egg protein).

Protein efficiency ratio (PER): weight gain per weight of protein eaten (usually measured for rats).

Introduction

The potato (*Solanum* spp.; Figure 1) is grown in 79% of the world's countries (FAO, 1986). It is second only to maize in terms of the number of producer countries and fourth after wheat, maize and rice in global tonnage. Its importance in European countries, the USSR, North America, Australia and the Andean countries of Latin America is well known. Less widely recognized, however, is the rapid growth rate of potato production in developing countries.

FAO statistics show that the percentage increase in potato production from 1961/65 to 1979, for all developing market economy countries, was greater than 99%, while that of cereals and other roots and tubers was, respectively, only 47% and 44% (International Potato Center, 1981). Potatoes are one of the most efficient crops for converting natural resources, labour and capital into a high quality food (Horton, 1981). They can yield more nutritious food material more quickly on less land and in harsher climates than most other major crops; and the edible food material can be harvested after only 60 days.

Current and future roles

Though potatoes occupy a smaller area in most developing countries than do other major food crops, their increasing popularity has caused planners and policy makers to take a closer look at the current and future roles that potatoes may play in national food production systems.

In recent years, the enormous potential for agronomic improvement in food plants through plant breeding has been increasingly recognized. In the case of potatoes, greater attention is being focused on ways to increase production, improve storage methods and facilitate marketing. Concomitantly, there is often a need to understand and improve the nutritional contributions that potatoes can make to the human diet. However, improvements in production and nutritional understanding may not go hand in hand, and one is often sacrificed to the other.

Many workers have advocated increasing the content and quality of protein in food crops, but this aspect is now often considered to be of minor importance. Although malnutrition is recognized as a deficiency in both energy and protein supplies, efforts to improve total food production have concentrated on breeding for greater crop yield and resistance to disease. This review emphasizes the importance of maintaining the good nutritional quality of the potato, while searching for the means to increase yields and enhance disease resistance.

Figure 1. Some of the varieties in the world potato collection of the International Potato Center.

Misconceptions and a remedy

There are many local misconceptions concerning the nutritional value of potatoes. In areas where it has been regarded as a luxury crop, the potato is often considered to be a ritual food or a garnish for other major meal components and, therefore to have aesthetic importance only (Figure 2). Where consumed as a complementary vegetable with staple food items, potatoes are often wrongly believed to make a negligible contribution to the nutritive value of a meal. Even where the potato is regarded as a staple food, it is usually seen only as an energy source and there is little awareness of its vitamin or protein content.

Obtaining factual information to correct these misunderstandings is difficult for government planners and workers within national potato programmes. Articles on potato nutritional quality are numerous, but most deal only with one specialized area of research. Reviews covering all aspects of the subject are extremely scarce and in most cases rather superficial. Furthermore, the journals in which either specialized or general articles appear are frequently unavailable in local libraries. Many articles address developed country interests, particularly those of the processing industry, and are therefore less relevant to the food needs of developing countries.

An outline of the book

This review is not intended to be fully comprehensive. Some aspects are mentioned only briefly since they are mainly the concerns of food scientists and technologists working in the large-scale processing industry. I have focused on the nutritional aspects of the potato in the form in which it is normally consumed in developing countries

Although the book is presented as a single unit, each chapter is written to stand alone and may be used in training courses or for other purposes. Not all the 700 titles reviewed (many of which are in the library at the International Potato Center) are cited; instead the references at the end of each chapter are those judged to be most useful or most current in respect of each chapter topic.

Chapter 1 introduces the reader to the structure and components of the potato; but does not cover all the morphology, structure and chemical reactions of the tuber in detail. These can be obtained in more specialized texts.

In *Chapter 2* the nutritional composition of the potato is compared with that of major food grains, roots, tubers and vegetables. Emphasis is on comparisons of foods, in the form in which they are usually eaten, i.e.

Figure 2. Potatoes appear as a garnish or flavouring in many meat and vegetable dishes offered by a market vendor in West Sumatra, Indonesia.

An outline of the book

cooked. Contributions the potato can make to dietary energy, protein, fibre, vitamin, mineral and trace element requirements of humans are discussed. This information is necessary for policy makers and planners making decisions about the allocation of research and development funds for food production; and ranks equal in importance with comparisons of yield, time to maturity, water, fertilizer and plant protection requirements, or perishability. However, in diets foods may be mixed to provide nutritious combinations suitable to local habits and conditions, and in this chapter we consider the potato as one component of such diets.

Potato nitrogen includes both protein and non-protein constituents. *Chapter 3* deals with factors affecting tuber nitrogen concentration and composition, and discusses feeding experiments with adults and children which demonstrate the high quality of potato N in human diets. The potential recovery of potato protein from processing waste is also reviewed.

The effects of cooking, processing and storage on the nutritive value of potatoes are considered in *Chapter 4*. Each process can alter significantly some components of the raw potato tuber, particularly the vitamins. The individual methods, however, have different effects, and those causing the least damage to nutritional quality are emphasized. Although the industrial processing methods mentioned may, at present, be of little relevance to most developing countries, some countries, notably India, are engaged in increasing the number of their processed-potato products.

Chapter 5 reviews current knowledge about potato glycoalkaloids, proteinase inhibitors and lectins, which may have adverse effects on humans. Relatively little is known about the nutritional significance of these components, particularly the last two. This chapter focuses on conditions producing increases in the concentrations of these components to toxic levels. The effects are discussed and current opinions on the nature of the toxic reactions are included. Practical recommendations for controlling tuber glycoalkaloid development are featured.

The International Potato Center has sponsored a three-year study to gain more knowledge of potato consumption in several developing countries. Some of the results are included in *Chapter 6*, to provide an overview of the variety of ways potatoes are used in human diets. Attention is directed to the nutritional contribution potatoes make within certain dietary patterns and their potential contribution if these patterns are altered or if current consumption levels are increased.

Finally, although the style of the book is conventionally English, 'french fries' and 'chips' are used throughout instead of the English terms chips and crisps, respectively. This is because it is felt that readers may be more familiar with the American usage. Also, energy values in both kcal and kJ are given throughout for the reader's convenience.

References

FAO (1986). *1985 FAO production yearbook*, vol. 36, pp. 126–7, FAO Statistics series no. 47. FAO, Rome.

Horton, D. (1981). A plea for the potato. *Ceres*, Jan–Feb: 2832.

International Potato Center (1981). *World potato facts*. International Potato Center, Lima.

1

Structure of the potato tuber and composition of tuber dry matter

Structure of the tuber

The potato tuber is not a root, but the enlarged apical portion of a lateral underground branch called a stolon. Externally, the tuber clearly shows its relation to the aerial stem. The spirally arranged eyes, from which sprouts arise, are formed by the base of a rudimentary leaf scale with three buds in its axil. The eyes may be shallow, medium or deep, depending on the plant variety. Figure 1.1 shows the organization of the principal internal tissues of the mature tuber.

The outer layer of single cells of the tuber apex, the epidermis, is usually colourless; but red and purple anthocyanin pigments are found in the periderm, which is a corky layer usually known as the skin.

Figure 1.1. Cross-section of the potato tuber showing internal structure.

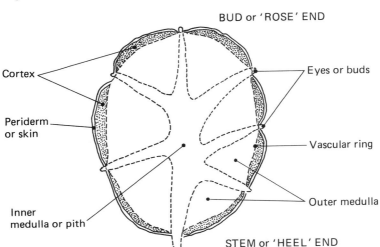

8   Tuber structure and dry matter

The region immediately inside the periderm extending inwards to the vascular ring is the cortical layer. This consists of two parts: that next to the skin is the cortex proper, usually not more than about 2 mm thick; between it and the vascular ring is a layer of storage parenchyma. The true cortex and the outer storage parenchyma are usually bracketed together as cortex. The total cortical layer is of variable thickness, usually 0.3 to 1 cm, but is negligible at the eyes and point of attachment. Inside the vascular ring is another layer of storage parenchyma called the outer medulla. The more translucent, wetter part in the centre of the potato is the inner medulla or pith. Table 1.1 gives the proportion of the whole potato in each major zone. Variations are considerable, owing to difficulties of defining the boundaries exactly and to differences between tubers.

The cell structure is relatively simple, consisting primarily of individual cells with cellulose walls cemented together with pectins. Inside a cell is the nucleus and the cytoplasm, which is the seat of processes such as respiration and starch synthesis. Some components such as starch grains are visible under the microscope as distinct inclusions in the cytoplasm; others are in solution in the cell sap.

Dry matter

Factors affecting the yield and content of potato tuber dry matter (DM) have been reviewed at length by Burton (1966) and more briefly by Grison (1973). DM content is extremely variable: a range of 13.7% to 34.8% was found amongst accessions to the germ plasm collection at the International Potato Center (unpublished data). Factors, other than variety, which influence DM include cultivation practices, climate, length of growing season, soil type, and pests and diseases.

Analyses of a number of varieties all show the same trend in distribution of DM: the percentage increases from the periphery inwards as far as the

Table 1.1. *Percentage contributions of the major parts of whole tubers*

Peel	Cortex	Outer medulla	Pith
5.9	33.0	35.4	25.7
2.8	52.2	31.3	13.7 (Small potatoes)
2.8	37.0	40.0	20.2 (Large potatoes)

Percentages as shown by Chappell (1958).

inner cortical tissue and outer medulla, and decreases from there to the centre. In other words, the bulk of the DM is contained in the storage parenchyma. In addition, there is a gradation of DM content from the 'heel', or attachment end, to the 'rose', or bud end, the former containing the higher percentage (Burton, 1966).

DM content is usually measured by specific gravity. The method is simple and rapid; Schéele et al. (1937) demonstrated a high correlation between DM and specific gravity when a large number of samples (560) was employed. However, the reliability of the relationship between specific gravity and total solids may be reduced when individual tubers containing intercellular air tissue spaces or the phenomenon known as 'hollow heart' are included in the measurements (Porter et al., 1964; Burton, 1966). Also, the regression lines calculated for the relationship can vary with factors such as soil type, growing conditions, location (Porter et al., 1964; Schippers, 1976), and even cultivars (Schippers, 1976). It has therefore been recommended that situation-specific regression lines be used.

The approximate composition of potato DM is given in Table 1.2, and

Table 1.2. *Approximate composition of potato tuber dry matter*[a]

Constituent	% in DM	
	Normal value (approx.)	Range (approx.)
Starch	70	60–80
Sucrose	0.5–1[b]	0.25–1.5[b]
Reducing sugars	0.5–2[b]	0.25–3[b]
Citric acid	2	0.5–7
Total N	1–2	1–2
Protein N	0.5–1	0.5–1
Fat	0.3–0.5	0.1–1
Fibre (dietary)[c]	6–8	3–8
Ash	4–6	4–6

[a] Source of all data, except those for dietary fibre, was Burton (1966).
[b] Figures represent mature, unstored tubers. Sugar content is affected by stage of maturity and temperature of storage. It is quite possible for total sugar content of potatoes stored at -1 °C to be 30% of DM and for that of unstored immature tubers to be more than 5% of DM (Burton, 1966).
[c] Calculated approximately from data given by authors mentioned in Chapter 2, pp. 38–9.

its various constituents will be described briefly. The nutritional significance will be discussed in Chapter 2.

Carbohydrates

Potato carbohydrates may be classified as starch, non-starch polysaccharides, and sugars.

Starch

Starch normally constitutes the greater part of the DM, and levels are modified by factors affecting the DM content of the tuber. Distribution of starch follows that of the DM, increasing from the skin inwards as far as the vascular ring and then decreasing inwards to the central medullary region, while the 'heel' end contains more starch than the 'rose' end.

Starch is present in the form of granules, consisting of amylopectin and amylose in a fairly constant ratio of 3:1. Amylopectin is a large, highly ramified molecule containing approximately 10^5 glucose residues. The amylose molecule is smaller, containing about 5000 glucose residues linked mainly by unbranched α-1,4 links, although slight branching sometimes occurs. There are small amounts of phosphorus, combined chemically with starch, most of it in the amylopectin fraction. The presence of phosphate appears to inhibit the complete enzymic breakdown of starch by β-amylase and affects the viscosity of gels prepared from potato starch (Burton, 1966).

When potatoes are subjected to heat in cooking or processing, the water they contain is absorbed into the starch granules, and at temperatures of 70 °C and above the starch is gelatinized. The resulting gel usually remains inside the potato cells unless these are ruptured during cooking or other processing treatments such as mashing, in which case the release of starch from the cells may make the cooked potato sticky.

This high starch content has made manufacture of potato starch economically feasible in developed countries. Such starch gels set rapidly and have a high hot-paste viscosity, unlike those from cereal starches. Potato starch is used in the manufacture of adhesives, in the textile industry, in the food industry and for the production of derived substances such as alcohol and glucose.

Non-starch polysaccharides

The non-starch polysaccharides comprise only a small part of the tuber DM: for example, Hoff & Castro (1969) found 5.6% on a dry weight basis (1.2% on a fresh weight basis) of cell wall–middle lamella material

Dry matter 11

in the tubers of the 'Superior' variety of *Solanum tuberosum*. Present in the cell wall and intercellular cementing substances of the middle lamella are cellulose, lignin, hemicelluloses, and pectin, which is mainly in an insoluble form but with some soluble calcium pectate. Hoff & Castro (1969) made a detailed study of the chemical composition of tuber cell wall material. Non-starch polysaccharides have an important role in the final texture of cooked potato. For example, during cooking the pectins are solubilized and to some extent degraded, causing separation, and occasionally rupture, of the cell walls. Classed together as dietary fibre, non-starch polysaccharides contribute to the nutritional value of the potato, as discussed in Chapter 2.

Starch and pectic substances as well as monovalent and divalent ions and cell size are amongst the factors most commonly cited as being responsible for controlling intercellular adhesion and hence the texture of cooked potato. This topic has been the subject of extensive studies, but the reasons for differences in texture characteristics of cooked potatoes and for texture defects are still under debate. The subject has been reviewed by Burton (1966) and Warren and Woodman (1974).

Sugars

Schwimmer *et al.* (1954) confirmed that sucrose, fructose and glucose are the major sugars in the white potato, although traces of some minor sugars were also found. Reducing sugars and sucrose occur in the tuber in only small amounts. They are of considerable importance, however, in the colour of products such as french fries (chips) and chips (crisps) and also play a part in potato flavour. The content of sugars in the potato tuber is influenced by variety, location and cultural treatment (Burton, 1966; Putz, 1976). In the mature stored tuber, sugars and starch exist in a state of dynamic equilibrium, which has been summarized by Burton (1966) as:

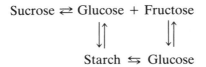

In newly harvested tubers at temperatures of 10 to 20 °C, 98% of the carbohydrate in this equilibrium are in the form of starch. However, respiration causes a perpetual turnover of tuber constituents, the equilibria among which are subject to change during storage. In particular, the content of sugars increases markedly at temperatures below 5 °C. For processed products, such as chips, french fries or home-cooked roast potatoes, a certain level of reducing sugars is essential for production of the popular golden-brown colour. This coloration is due to the non-enzymic browning, or 'Maillard', reaction in which sugars react with amino acids, ascorbic acid, and other organic compounds, to produce brown pigments known as 'melanoidins'. However, excessive production of sugar from starch leads to what is generally regarded as an unpleasant sweet flavour in cooked potatoes. Sugars are reconverted to starch if potatoes, previously stored at lower temperatures are held, for two to four weeks, at 10 to 20 °C – a process known as 'reconditioning'. Isherwood (1976) gives details of the mechanism of starch–sugar interconversion during storage of *S. tuberosum*, while tubers were transferred from 10 °C to 2 °C and back to 10 °C. The accumulation of sugars during prolonged storage at higher temperatures (particularly 10 to 20 °C) is known as senescent sweetening and is irreversible, probably because of cell membrane breakdown. It is interesting to note that farmers in the Mantaro Valley of Peru prefer potatoes sweetened by long-term home storage (Werge, 1980).

Nitrogen
Protein N (with the exception of enzymes, which are dealt with briefly in the present chapter) and non-protein N are fully described in Chapter 3. The glycoalkaloids, proteinase inhibitors and lectins, which make up part of the basic tuber N, are discussed in Chapter 5.

Lipids
The lipid content of potato is low. Galliard (1973) found 0.08 to 0.13% (FWB) in 23 varieties. This range is too low to have any nutritional significance but contributes towards potato palatability (Kiryukhin & Gurov, 1980), enhances tuber cellular integrity and resistance to bruising and plays a part in reducing enzymic darkening in tuber flesh (Mondy & Mueller, 1977). The greater importance of the lipids lies in their susceptibility to enzymic degradation and non-enzymic auto-oxidation, which cause 'off' flavour and rancidity in dehydrated and 'instant' potato products. Galliard (1973) has described the fatty acid composition of the lipid, and Galliard & Matthew (1973) have studied the lipid-degrading enzymes in 23 varieties, in relation to the production of potatoes for processing. Approximately 75% of the total fatty acids of the lipids are the polyunsaturated linoleic and linolenic acids. These contribute to production of both desirable flavour characteristics in cooked tubers and undesirable 'off' flavours in processed products. The polyunsaturated acids are rapidly converted to free fatty acids and other compounds by

lipid-degrading enzymes during tuber processing and are also extremely susceptible to auto-oxidation.

Enzymes

The tuber contains numerous enzyme systems that constitute a considerable proportion of the total protein, and some of which are therefore important in the potato as a food source. Their various functions have been described by Burton (1966), and reviewed more recently by Smith (1977).

According to Burton (1978), the mechanism of low-temperature sweetening during storage has been explained in terms of the relative activities of enzymes and enzyme inhibitors. Phosphorylase (the action of which may be reduced by an inhibitor at higher temperatures) breaks down starch to glucose-1-phosphate. Some of this is converted to sucrose by sucrose phosphate synthetase. Subsequent hydrolysis of part of the sucrose to glucose and fructose is regulated by the comparative levels of β-fructo-furanosidase (invertase) and a β-fructo-furanosidase inhibitor.

Of the lipid-degrading enzymes, one is a lipolytic acyl hydrolase that liberates free fatty acids from phospholipids and glycolipids and the other is a lipoxygenase that converts linoleic and linolenic acids to their 9-hydroperoxide derivatives (Galliard & Matthew, 1973). The importance of these enzymes in food processing is their probable involvement in the formation of volatile flavour and 'off' flavour compounds, as mentioned above.

A further enzyme system, important during the preparation of both home-cooked and industrially processed potatoes, is that causing enzymic discoloration or blackening of peeled or cut tubers. When tuber cells are injured, polyphenoloxidase (tyrosinase) gains access to tyrosine and other orthodihydric phenols, which are thus oxidized to dark or black compounds (melanins). Since this reaction is initiated when cells are injured, some purchased potatoes may already have enzymic blackening of part of the flesh, because of rough handling or mechanical damage. This results in wastage when tubers are prepared for cooking (or after cooking in skins), since the aesthetically displeasing blackened parts will normally be discarded. Similarly, high rates of enzymic darkening are undesirable in industrial processing especially for producers of pre-peeled potatoes. Matheis & Belitz (1977a,b,c, 1978) conducted a series of studies on enzymic browning of potatoes including the reactants involved, and the kinetics of the reaction. Burton (1966) briefly reviewed factors affecting susceptibility of potatoes to enzymic discoloration. These include variety, cultural practices and climatic conditions.

Methods of inhibition of enzymic discoloration caused by peeling have been reviewed by Smith (1977).

Organic acids

The major organic acids identified in the potato are citric and malic acids (Jadhav & Andrew, 1977; Bushway *et al.*, 1984). Others present are oxalic and fumaric (Bushway *et al.*, 1984), chlorogenic and phosphoric (Schwartz *et al.*, 1962), as well as ascorbic, nicotinic and phytic acids, amino acids and fatty acids. All these contribute to flavour and help to buffer the potato sap (the pH of the tuber is 5.6 to 6.2); the level of some, especially that of malic acid, can be used to indicate tuber maturity. Ascorbic and nicotinic acids influence directly, and phytic acid indirectly, tuber nutritional value (see pp. 45 and 49).

Chlorogenic acid can react with ferric ions on cooking to produce a dark-coloured complex. This phenomenon, known as post-cooking or non-enzymic blackening, may occur more in the 'heel' than in the 'rose' end of potatoes, in which case it is called stem-end blackening. Citric acid in tubers helps to prevent this by sequestering the iron present and making it unavailable for forming a complex with chlorogenic acid. Susceptibility of potatoes to post-cooking blackening therefore depends upon the relative concentrations of iron and of chlorogenic and citric acids in the tubers (Burton, 1966).

Kyle (1976) has reviewed all types of discoloration in potatoes, including enzymic browning and non-enzymic Maillard browning.

Pigments

Anthocyanin pigments in the periderm and peripheral cortex produce totally or partly pigmented skins in potatoes. In some South American varieties, the pigment is so dark that tubers may appear black and others dark purple.

Potato flesh may be white or various shades of yellow, depending upon the variety. Yellow coloration is generally due to presence of carotenoid pigments. The major carotenoid identified in 13 German varieties was violaxanthin, followed by lutein and lutein-5,6-epoxide and, in lower concentrations, by neoxanthin A and neoxanthin (Iwanzik *et al.*, 1983); β-carotene was detected in only trace amounts or was totally absent. One cultivar had an intense yellow flesh colour, but a relatively low level of carotenoids. It is therefore possible that, in some varieties, the yellow colour is due to other, unidentified, pigments as well as to carotenoids. In some places (e.g. Peru) yellow-fleshed varieties are highly prized and command higher prices than those with white flesh.

Dry matter

When harvested, potato tubers are exposed to light, and chlorophyll forms in the superficial parts, giving a green colour. The association of this chlorophyll with an increase in tuber glycoalkaloid content is discussed in Chapter 5.

Vitamins

Potatoes are substantial sources of several vitamins: ascorbic acid (vitamin C) and the B vitamins thiamin (B_1), pyridoxine (B_6) and niacin. Riboflavin (B_2), folic acid and pantothenic acid are also present. Small amounts of vitamin E have been reported (Paul & Southgate, 1978). Biotin is present in traces. The vitamin A precursor β-carotene is absent or present only in trace amounts. The ranges of content of various significant vitamins, the factors affecting variation in their concentration and their nutritional contributions are reviewed in Chapter 2.

Vitamin C exists in the tuber in both the reduced and oxidized forms (see Figure 1.2). In the freshly harvested raw tuber the reduced form, L-ascorbic acid, is quantitatively the most important (Lampitt et al., 1945; Leichsenring et al., 1957) or is the only form present (Wills et al., 1984). The enediol group on C-2 and C-3 of this compound can be oxidized to a diketo group in a reversible reaction. The resulting dehydroascorbic acid has full vitamin C activity. However, dehydroascorbic acid is more labile than the reduced form and is readily, and irreversibly, oxidized to 2,3-diketogulonic acid, with consequent loss of vitamin C activity. These oxidations may occur during post-harvest operations.

Determinations of vitamin C content in stored, cooked or processed

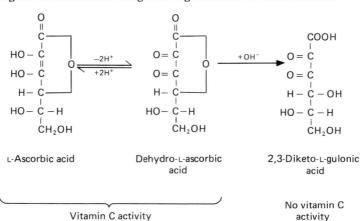

Figure 1.2. Structural changes during oxidation of L-ascorbic acid.

potatoes use methods which measure (1) L-ascorbic acid only, or (2) the total of L-ascorbic and dehydroascorbic acids. The former methods can give misleadingly low values for vitamin C concentrations, or conversely can imply misleading high losses of vitamin C. Where possible, the text specifies which form of vitamin C was reported in tuber analyses. There is further discussion of the changes in different forms of vitamin C which result from post-harvest handling in Chapter 4.

Ash

Ash (non-volatile inorganic residue) constitutes about 1% of the tuber fresh weight (about 4% to 6% of the DM) and contains a large variety of minerals and trace elements. Some of these have a function in the life of the plant; others may be present only because they were in solution in the soil in which the tubers were grown, or were derived from fertilizers or sprays employed during cultivation. Potatoes are rich in potassium and phosphorus but are rather poor sources of sodium and calcium. Iron is always found as are magnesium, zinc, manganese, silicon, sulphur and chlorine. There are traces of boron, copper, iodine, bromine, aluminium, arsenic, molybdenum, cobalt and nickel (Lampitt & Goldenberg, 1940).

Concluding remarks

This brief review of the structure and composition of the potato tuber is intended to introduce the major components, some of which will be discussed in more detail from a nutritional viewpoint in Chapter 2.

Discoloration of potatoes by enzymic and non-enzymic browning, the interconversion of starch and sugars and the production of 'off' flavours from potato lipids have been dealt with only superficially. However, it is beyond the scope of the book to treat at length various topics mentioned in this chapter, and the cited sources will provide interested readers with more detailed information.

References

Burton, W. G. (1966). *The potato*, 2nd edn. Drukkerij Veenman BV, Wageningen.

Burton, W. G. (1978). Post-harvest behaviour and storage of potatoes. In Coaker, T. H. (ed.), *Applied Biology*, vol. 2. Academic Press, New York.

Bushway, R. J., Bureau, J. L. & Mcgann, D. F. (1984). Determinations of organic acids in potatoes by high performance liquid chromatography. *J. Food Sci.* **49**: 75–7, 81.

Chappell, G. M. (1958). *The potato as a food*. Potato Marketing Board, London.

Galliard, T. (1973). Lipids of potato tubers. 1. Lipid and fatty acid composition of tubers from different varieties of potato. *J. Sci. Food Agric.* **24**: 617–22.

Galliard, T. & Matthew, J. A. (1973). Lipids of potato tubers. 2. Lipid-degrading enzymes in different varieties of potato tuber. *J. Sci. Food Agric.* **24**: 623–7.

Grison, C. (1973). [The dry matter.] In French. Information fiche no. 53. Institut Technique de la Pomme de Terre, Paris.

Hoff, J. E. & Castro, M. D. (1969). Chemical composition of potato cell wall. *J. Agric. Food Chem.* **17**: 1328–31.

Isherwood, F. A. (1976). Mechanism of starch–sugar interconversion in *S. tuberosum*. *Phytochemistry* **15**, 33–41.

Iwanzik, W., Tevini, M., Stute, R. & Hilbert, R. (1983). [Carotenoid content and composition of various German potato varieties and its relationship to tuber flesh colour.] In German. *Potato Res.* **26**: 149–62.

Jadhav, S. J. & Andrew, W. T. (1977). Effects of cultivars and fertilizers on non-volatile organic acids in potato tubers. *Can. Inst. Food Sci. Technol. J.* **10**: 13–18.

Kiryukhin, V. P. & Gurov, V. A. (1980). [Fatty acids and palatability of potatoes.] In Russian. *Kartofel Ovoshchi* no. 5, 10–11.

Kyle, W. S. A. (1976). Discolouration of potatoes and potato products. *Proc. Inst. Food Sci. Technol.* **9**: 93–8.

Lampitt, L. H. & Goldenberg, N. (1940). The composition of the potato. *Chem. Ind.* **59**: 748–61.

Lampitt, L. H., Baker, L. C. & Parkinson, T. L. (1945). Vitamin C content of potatoes. II. The effect of variety, soil and storage. *J. Soc. Chem. Ind.* **64**: 22–6.

Leichsenring, J. M., Norris, L. M. & Pilcher, H. L. (1957). Ascorbic acid contents of potatoes. I. Effect of storage and of boiling on the ascorbic, dehydroascorbic and diketogulonic acid contents of potatoes. *Food Res.* **22**: 37–43.

Matheis, G. & Belitz, H. D. (1977*a*). [Studies on enzymic browning of potatoes (*Solanum tuberosum*). I. Phenoloxidases and phenolic compounds in different varieties.] In German. *Z. Lebensm.-Unters. Forsch.* **163**: 92–5.

Matheis, G. & Belitz, H. D. (1977*b*). [Studies on enzymic browning of potatoes. 2. The quantitative relationship between browning and its causative factors.] In German. *Z. Lebensm.-Unters. Forsch.* **163**, 186–90.

Matheis, G. & Belitz, H. D. (1977*c*). [Studies on enzymic browning of potatoes. 3. Kinetics of potato phenoxidase (E.C. 1.14.18.1 monophenol, dihydroxyphenylalanine: oxygen-oxido reductase).] In German. *Z. Lebensm.-Unters. Forsch.* **163**: 191–5.

Matheis, G. & Belitz, H. D. (1978). [Studies on enzymic browning of potatoes. 4. Relationship between tyrosine turnover and rate of browning.] In German. *Z. Lebensm.-Unters. Forsch.* **167**: 97–100.

Mondy, N. I. & Mueller, T. O. (1977). Potato discoloration in relation to anatomy and lipid composition. *J. Food Sci.* **42**: 14–18.

Paul, A. A. & Southgate, D. A. T. (1978). *McCance and Widdowson's, The composition of foods*, 4th edn, MRC Special Report no. 297. HMSO, London.

Porter, W. L., Fitzpatrick, T. J. & Talley, E. A. (1964). Studies of the relationship of specific gravity to total solids of potatoes. *Am. Potato J.* **41**: 329–36.

Putz, B. (1976). [Effect of cultivation procedures on the sugar content of potato tubers. I. Effect of year, growing site and variety.] In German. *Kartoffelbau* **27**, 230–1.

Schéele, C. von, Svensson, G. & Rasmusson, J. (1937). [Determination of starch content and dry matter of the potato by means of specific gravity]. In German. *Landw. VersSta.* **127**: 67–96.

Schippers, P. A. (1976). The relationship between specific gravity and percentage dry matter in potato tubers. *Am. Potato J.* **53**: 111–22.

Schwartz, J. H., Greenspun, R. B. & Porter, W. L. (1962). Identification and determination of the major acids of the white potato. *Agric. Food Chem.* **10**: 43–6.

Schwimmer, S., Bevenue, A., Weston, W. J. & Porter, A. L. (1954). Survey of major and minor sugar and starch components of the white potato. *Agric. Food Chem.* **2**, 1284–90.

Smith, O. (1977). *Potatoes: production, storing, processing*, 2nd edn. AVI Publishing Co. Inc. Westport CT.

Warren, D. S. & Woodman, J. S. (1974). The texture of cooked potatoes: a review. *J. Sci. Food Agric.* **25**: 129–38.

Werge, R. W. (1980). *Potato storage systems in the Mantaro valley region of Peru*. International Potato Center, Lima.

Wills, R. B. H., Wimalasiri, P. & Greenfield, H. (1984). Dehydroascorbic acid levels in fresh fruit and vegetables in relation to total vitamin C activity. *J. Agric. Food Chem.* **32**: 836–8.

2

The nutritional value of the components of the tuber

Potatoes are thought to have originated in the Andean highlands of South America (Figure 2.1), and for thousands of years they have been used to maintain and support the growth and health of large numbers of humans. Salaman (1949) asserts that, through discovery of the potato, the ensuing cultivation of frost-hardy types, and the development of methods of preserving tubers, man was able to live at greater altitudes and thus gain mastery of the continent of South America. The dependence of the Irish and Scots on potatoes as their principal source of nourishment and the resulting famine in 1846–47, when the potato crop failed due to blight, are well documented (Woodham-Smith, 1962; Salaman, 1949). Anthropologist Fürer-Haimendorf (1964) has argued that the introduction of the potato into the Sherpa Khumbu region of Nepal stimulated population growth and provided the agricultural surplus necessary for the rise of the elaborate Buddhist civilization in the northern part of the country.

However, although the potato has been shown to be a source of good-quality protein, to have a favourable ratio of protein calories to total calories and to be an important source of vitamins and minerals, its overall value in the diet nowadays is generally greatly underestimated. This chapter demonstrates the value of potato, particularly for consumers in developing countries, where diets are principally made up of basic regional foods. Data are from nutrition studies, interpretations of experimental results and discussion of the nutritional quality of the potato relative to known requirements or recommended levels of nutrient intake, drawn from WHO (1985), FAO/WHO (Passmore *et al.*, 1974) and from the US National Research Council (1980).

One practical goal of food and nutrition policy planners in developing countries is to reduce disparities between requirements and intakes of nutrients. Comparison of the nutrient contents of various food crops

20 *Nutritional value of the components*

indicates which foods may be of particular value in improving local diets and therefore should have increased production. This chapter provides such comparisons between the potato and other crops, which can be used in a variety of situations. The potato is a staple food in those tropical regions where elevation provides a moderate cool-season climate (Figure 2.2). In other tropical areas it is used as a vegetable (Figure 2.3). It is therefore compared nutritionally with root and tuber staples, with cereal grains and with dry beans, as well as with several vegetable items; emphasis is on comparisons of the composition of foods in the form in which they are eaten.

Figure 2.1. A pre-Columbian ceramic pot in the form of a potato tuber.

Nutritional value of the components 21

Chemical compositions refer to the components of the flesh only or of the whole potato, and values given in the tables are for average composition. Contents of all components may vary considerably and depend upon, for example, variety, location of growth, soil type, climate and conditions of cultivation. Composition is altered by storage and by cooking or processing and is also subject to the different methods of sampling and analysis used by different investigators. These points were

Figure 2.2. Potatoes retailed as a staple in Zaïre.

Figure 2.3. Retailing potatoes as a vegetable in Indonesia (above) and Pakistan (below).

Dry matter

A sample of 100 g of potato contains, on average, about 20 g of DM, of which only about 0.1 g is in the form of lipid. Quantitatively, the major constituent is starch (Table 2.1).

DM content is, however, extremely variable. The interactive factors (variety, climate and soil conditions, agricultural practices, length of growing season, incidence of pests and diseases) which influence tuber DM have been reviewed at length by Burton (1966).

Gross amounts of energy and protein furnished by the potato are dependent upon its DM content and the latter is hence of considerable significance to the potato consumer.

By utilizing figures for total N (*fresh weight* basis) and DM of 328 North American clone samples and 627 samples from the International Potato Center's germplasm collection in Peru, it has been shown that total N and DM are positively correlated (Poats & Woolfe, 1982). There may be confusion over improvements in nutritional quality if only the negative correlation between total N (on a *dry weight* basis) and DM is reported (Poats & Woolfe, 1982). This indicates the need for researchers to express results on a basis meaningful to the needs of potato consumers. The positive relationship between tuber nutritive value and DM should encourage plant breeders to maintain DM levels when searching for increased yields and resistance to pests and diseases in new varieties.

Table 2.1. *Average content of major constituents of the potato tuber*

Constituent	Weight (% of total tuber)
Water	80
Dry matter:	20
Carbohydrate	16.9
Protein	2.0
Lipid	0.1
Ash	1.0

Energy and protein
Energy value

The potato has a lower average carbohydrate content than do other roots and tubers, and also a comparable fat content (Table 2.2). Raw potato has a somewhat lower average energy content than other raw roots and tubers with 335 kJ (80 kcal) per 100 g. However, the large variation in tuber DM content produces a range of energy contents also, e.g. 264 to 444 kJ (63 to 106 kcal) per 100 g was found for the energy values of North American commercial varieties (Toma et al., 1978a). The energy content of raw potato is considerably less than that of raw cereals and legumes; however, when cooked, the latter staples absorb large quantities of water, which changes their composition significantly. The potato, when boiled in its skin, retains its energy value almost unaltered. A fairer comparison of the potato and the cereals or legumes, therefore, is either on a dry, raw basis or on a cooked, 'as eaten', basis. Doughty

Table 2.2. *Composition of raw potatoes and other roots and tubers (per 100 g edible portion)*[a]

Food	Energy (kJ)	Energy (kcal)	Moisture (%)	Crude protein (g)	Fat (g)	Total carbo-hydrate (g)	Dietary fibre (Crude fibre) (g)
Potato (*Solanum tuberosum*)	335	80	78.0	2.1	0.1	18.5	2.1[b] (0.5)
Sweet potato (*Ipomoea batatas*)	485	116	70.2	1.4	0.4	27.4	2.5[b] (0.9)
Yam (*Dioscorea* spp.)	444	106	72.0	2.2	0.2	24.2	4.1[b] (0.7)
Cocoyam. taro (*Colocasia* spp.)	423	101	73.2	1.9	0.2	23.5	(0.8)
Cassava (*Manihot esculenta*)	607	145	62.6	1.1	0.3	35.2	5.2[g] (1.0)

[a] Averages of figures from: Wu Leung & Flores (1961); Wu Leung et al. (1968); Wu Leung et al. (1978); Caribbean Food & Nutrition Inst. (1974); Watt & Merrill (1975).
[b] Paul & Southgate (1978).
[c] See also recent values given in Table 2.11.
[d] Pale variety.
[e] Yellow variety.
[f] Deep yellow variety.
[g] Nyman et al. (1984).

Energy and protein

(1982) has rightly called water the 'hidden ingredient' in foods. When calculated on the basis of a moisture content equivalent to that of the dry staples, the energy content of the potato is similar to that of cereals or *Phaseolus* beans (Table 2.3). Table 2.4, on the other hand, compares cooked potatoes with other cooked staples and shows that, while the potato is still lower in energy value than most of the other cooked sources, the differences are not as marked as those observed on a raw basis. Bread and tortillas, however, provide substantially more energy than cooked potatoes.

The potato's low energy density (energy content per gram of food) is advantageous when potatoes are included (without added fat or energy-rich sauces) in diets of the developed world, where obesity, as a state of malnutrition, is found increasingly. In parts of the developing world where diets are energy deficient, this attribute may be a disadvantage, particularly in the diet of infants and small children, whose digestive

Ash (g)	Ca (mg)	P (mg)	Fe (mg)	β-carotene equivalents (μg)	Thiamin (mg)	Ribo-flavin (mg)	Niacin (mg)	Ascorbic acid (mg)
1.0	9	50	0.8	0–trace	0.10^c	0.04^c	1.5^c	20
0.8	33	46	1.1	47^d 1468^e 2108^f	0.11	0.05	0.7	26
1.0	25	53	0.9	Trace–10	0.10	0.03	0.5	9
1.2	38	75	1.2	Trace	0.13	0.03	0.9	6
0.9	38	41	1.0	0–30	0.06	0.04	0.6	36

systems cannot cope with large intakes. Too much potato would be needed to supply all the energy requirements of small children, so they need an energy-rich supplement. However, they can consume as much as 50% to 75% of their energy supply as potato (López de Romaña et al., 1981). Although adults would also have to consume large quantities to meet all their daily energy needs, their digestive systems have a greater capacity. Up to 4.5 kg per capita were consumed daily in Ireland in the seventeenth to nineteenth century (Pimental et al., 1975). This would

Table 2.3. *Composition of raw potato, dried potato and other plant foods (per 100 g edible portion)*[a]

Food	Energy (kJ)	Energy (kcal)	Moisture (%)	Crude protein (g)	Fat (g)	Total carbo-hydrate (g)	Dietary fibre (Crude fibre) (g)
Potatoes	335	80	78.0	2.1	0.1	18.5	2.1[b] (0.5)
Potato (dried)[d]	1343	321	11.7	8.4	0.4	74.3	8.4[e] (2.0)
Plantain	531	127	64.5	1.2	0.2	33.3	5.8[b] (0.5)
Corn, sweet	402	96	72.7	3.5	1.0	22.1	3.7[b] (0.7)
Corn, mature	540	129	63.5	4.1	1.3	30.3	(1.0)
Corn, dried	1498	358	11.5	9.5	4.4	73.2	9.3[f] (2.1)
Rice, milled white	1523	364	12.0	6.8	0.5	80.2	2.4[b] (0.4)
Wheat, hard	1389	332	12.3	13.3	2.0	70.9	12.1[f] (2.3)
Sorghum whole grain	1431	342	10.9	10.1	3.4	73.2	9.0[f] (2.0)
Beans (*Phaseolus vulgaris*) dry	1414	338	11.8	21.9	1.6	61.2	25.4[b] (4.4)

[a] Averages of figures from references in the footnotes to Table 2.2.
[b] Paul & Southgate (1978).
[c] See also Table 2.11.
[d] Potato theoretically 'dried' to a moisture content of 11.7%, which is the average of moisture contents of dry foods shown in the table.
[e] Calculated from Paul & Southgate (1978).
[f] Nyman et al. (1984).
[g] White variety.
[h] Yellow variety.

Energy and protein

have provided approximately 15.06 MJ (3600 kcal) and 94 g of total protein.

Starch furnishes most of the energy supplied by the potato. Digestibility of this starch influences the energy value of the potato and hence also the bulk of potato which must be eaten to supply a given amount of energy. The digestibility of potato starch, low in the raw state, is greatly improved during cooking or processing, but there is some doubt about the extent of digestibility of cooked potato starch in infants and small children (see Chapter 4). This aspect deserves further investigation.

Ash (g)	Ca (mg)	P (mg)	Fe (mg)	β-carotene equivalents (μg)	Thiamin (mg)	Riboflavin (mg)	Niacin (mg)	Ascorbic acid (mg)
1.0	9	50	0.8	0–trace	0.10^c	0.04^c	1.5^c	20
4.0	36	201	3.2	Trace	0.40	0.16	6.0	80
1.0	9	350	0.9	125–780	0.08	0.04	0.6	22
0.7	3	111	0.7	400	0.15	0.12	1.7	12
0.8	5	128	1.1	35	0.18	0.08	1.9	9
1.3	12	251	3.4	6^g 147^h	0.35	0.11	1.9	Trace
0.6	20	115	1.1	0	0.08	0.04	1.8	0
1.7	44	359	3.9	0	0.52	0.12	4.4	0
1.7	32	290	4.9	0–20	0.39	0.15	3.8	0
3.7	98	247	7.6	0–20	0.53	0.19	2.2	Trace–3

Table 2.4. *Composition of cooked potatoes and other cooked plant foods (per 100 g edible portion)*[a]

Food	Energy (kJ)	Energy (kcal)	Moisture (%)	Crude protein (g)	Fat (g)	Total carbo-hydrate (g)	Dietary fibre (Crude fibre) (g)
Potato (boiled in skin, flesh only)	318	76	79.8	2.1	0.1	18.5	1.07[b] 1.30[c] (0.5)
Cassava (boiled)	519	124	68.5	0.9	0.1	29.9	(0.4)
Plantain (green, boiled)	393	94	74.5	1.1	0.2	23.8	6.4[d] (0.2)
Rice (boiled, white)	565	135	67.9	2.3	0.3	28.0	0.8[d] (0.1)
Maize (porridge)	318	76	81.2	1.8	0.8	15.6	(0.2)
Tortilla (lime-treated maize)	879	210	47.5	4.6	1.8	45.3	(0.8)
Bread (un-enriched white)	1163	278	32.7	8.7	1.6	55.7	2.7[d] (0.5)
Macaroni or spaghetti (boiled)	552	132	66.0	4.1	0.7	26.7	(0.2)
Sorghum (porridge)	356	85	79.6	2.7	0.5	17.0	(0.2)
Beans (*Phaseolus vulgaris*, boiled)	494	118	69.0	7.8	0.5	21.4	7.4[d] (1.5)

[a] Averages of figures from references in footnotes to Table 2.2.
[b] Boiled, peeled. Average of figures from Jones *et al.* (1985).
[c] Boiled with skin, including skin. Average of figures from Jones *et al.* (1985).
[d] Paul & Southgate (1978).

Ash (g)	Ca (mg)	P (mg)	Fe (mg)	β-carotene equivalents (μg)	Thiamin (mg)	Riboflavin (mg)	Niacin (mg)	Ascorbic acid (mg)
0.9	7	53	0.6	0-trace	0.09	0.03	1.5	16
0.6	—	—	—	—	—	—	—	26
0.6	9	32	1.2	345	0.04	0.06	0.6	12
0.7	8	36	0.3	0	0.02	0.01	0.4	0
0.6	4		0.6	0	0.06	0.01	0.5	0
0.8	196	138	2.6	20	0.15	0.05	1.0	0
1.4	24	98	1.3	0	0.09	0.05	1.0	0
2.4	8	59	0.5	0	0.01	0.01	0.3	0
0.2	4	31	1.7	0	0.04	0.01	0.2	0
1.3	38	140	2.4	Trace	0.11	0.06	0.7	0

Protein content

Potato crude protein (total N × 6.25) at about 2% (FWB) is comparable to that of most other root and tuber staples, with the exception of cassava, which has only half this amount (Table 2.2). It is also comparable, on a dry basis, with that of the cereals (Table 2.3) and, on a cooked basis, with that of boiled rice or of cereals cooked as porridges (Table 2.4).

Factors influencing the level of total N are reviewed in Chapter 3, which also discusses the high nutritional value of potato N. Table 2.5 shows the essential amino acid composition of potato in comparison with four other important staples. The advantage the potato has over the cereal staples is its high lysine content. However, it contains lower concentrations of the sulphur-containing amino acids (methionine, cystine/cysteine) than the cereals. *Phaseolus* beans also have greater quantities of lysine but have even lower amounts of sulphur amino acids. In combination with other foods, potatoes can supplement diets which are limiting in lysine, e.g. rice with accompanying potatoes provides a better quality protein. In some developing countries, meals are frequently served with mixtures of boiled potato and rice and/or pasta. Such mixtures, however, are often wrongly assumed by developed-country consumers to provide nothing more than large quantities of energy.

It has been suggested (Werge, 1979) that the comparative advantage of the potato as a foodstuff in the tropics, on a unit weight basis, lies in its

Table 2.5. *Essential amino acid composition of potatoes and four other plant foods*[a]

Food	Essential amino acids (g/16 g N)								
	His	Ile	Leu	Lys	Met + Cys	Phe + Tyr	Thr	Trp	Val
Potato	1.9	4.2	6.1	5.4	2.9	7.4	3.8	1.4	5.1
Wheat (white flour)	2.1	3.8	7.0	1.9	4.2	7.4	2.7	1.1	4.3
Rice (milled, white)	2.4	3.8	8.2	3.7	3.7	8.8	3.4	1.3	5.8
Oatmeal	2.1	3.8	7.2	3.7	4.5	8.3	3.4	1.3	5.1
Beans (*Phaseolus vulgaris*)	2.9	4.2	7.7	7.2	1.9	7.9	4.0	1.0	4.6

[a] Calculated from data by Paul & Southgate (1978).

ability to supply high-quality protein. Using the latest figures for energy and protein requirements (WHO, 1985), it can be calculated that 100 g (one small tuber) of potato can supply 7%, 6% and 5% of the daily energy and 12%, 11% and 10% of the daily protein needs of children aged 1–2, 2–3 and 3–5 years, respectively. For adults, depending upon body weight and sex, 100 g of potato can supply 3% to 6% of daily protein requirements. Even in a country as affluent as Britain, potato contributed 3.4% of the total household protein intake in 1983 (National Food Survey Committee, 1983). This can be compared with the contributions of fruit (1.3%), eggs (4.6%), fish (4.8%), cheese (5.8%), beef (5.7%), white bread (9.8%) and milk (14.6%).

For infants and children, an energetically adequate diet cannot support growth if its protein content is inferior to requirements. Conversely, with an energetically inadequate diet, protein is metabolized as an energy source, instead of being used for growth. It is therefore essential to consider the quality of a food or diet in terms of both protein and energy: for example, by the percentage of the total energy supplied by protein. A useful estimate of protein quality is the net dietary protein calories percentage or NDpCal%, which also utilizes the chemical or protein score, since many vegetable foods are limiting in at least one of the essential amino acids. NDpCal% is calculated as:

$$\frac{(\text{g protein}/100 \text{ g food}) \times \text{chemical score} \times 4}{\text{total kcal}/100 \text{ g food}} \times 100$$

The NDpCal% of breast milk is approximately 8, which meets the known requirements of infants. As children grow, energy requirements increase and the percentage drawn from protein decreases until, at one year old, the child requires an NDpCal% of about 6. Table 2.6 shows the NDpCal% of some staple foods and indicates that potato and some cereals have approximately the correct balance between net protein calories and total calories for the end of the first year of life and are adequate for all other age groups. Potato is, therefore, a well-balanced food in terms of protein and energy. In other words, if enough is eaten to supply energy needs, a significant amount of protein will also be provided.

NDpCal% is also a means of estimating the quantity of a food or diet which is needed to meet the protein and energy requirements of a section of the population. Table 2.7 shows the quantities of relatively expensive protein-rich supplements that should be added to the more economical staples to give mixes which provide adequate energy and protein for the infant or young child, while minimizing bulk. Each mix provides approximately 1464 kJ (350 kcal) and has an NDpCal% of 6 when oil and/or sugar are added. Regional food availabilities and costs, food preferences, preparation times and cooking fuel costs might also be considered on a local basis when determining the relative utility of the mixes. Table 2.7 was compiled as a practical guide to infant feeding (Cameron & Hofvander, 1976). It can also be used, in the present context, to compare potatoes with other staples, showing that:

(1) Potato requires less of the expensive protein-rich supplements than do other root, tuber and fruit staples, due to the high quality of its protein.
(2) Potato requires the addition of quantities of supplements similar to those combined with cereals, with the exception of oats, which requires less of all the supplements.
(3) Potato is the staple required in greatest quantity on a raw basis, due to its low energy density. On a cooked basis, the disparity between quantities of cereal or potato is reduced, or in the case of oats, reversed. This point could be considered when the supplement is to be added to a portion of the cooked staple removed from the family pot.

Table 2.6. *NDpCal% of potatoes and other staple foods compared to that of breast milk*[a]

Food	NDpCal%[b]
Breast milk	8.0
Oats	7.0
Potatoes	6.0
Wheat	6.0
Sorghum	4.9
Rice	4.9
Yam	4.8
Maize	4.5
Sweet potato	3.4
Plantain	1.5
Cassava	<1

[a] Data from Cameron & Hofvander (1976).
[b] NDpCal% = Net dietary protein calories percentage or the percentage of total energy provided by protein. (Note: Requirement for adult = 4.0; requirement for a one-year old child = 6.0.)

Table 2.7. Staple–supplement combinations for infant or child feeding[a]

Supplements (g) \ Staple (g)	Oats		Wheat		Rice		Millet or sorghum		Maize		Potato		Sweet potato		Yam		Taro		Plantain		Cassava flour	
Egg[b]	65	10	65	25	65	30	65	30	65	25	300	25	180	35	220	25	190	25	150	45	60	50
Dried skim milk[b]	65	5	65	10	65	15	60	15	60	15	280	15	175	20	190	15	180	15	150	20	60	20
Dried whole milk[b]	55	10	55	15	45	25	45	20	40	25	220	20	100	30	115	30	115	25	90	30	35	30
Fish[b]	65	15	70	30	70	30	70	25	70	20	310	25	210	35	240	35	220	40	180	45	75	50
Chicken lean meat[b]	65	10	65	20	65	25	65	25	65	35	300	25	180	35	210	35	195	30	160	45	70	45
Beans[c]	75	5	80	10	65	25	75	10	55	35	320	20	125	50	165	40	150	45	85	55	40	55

[a] Data from Cameron & Hofvander (1976). Combinations formulated to minimize bulk, and to provide 1464 kJ (350 kcal) with an NDpCal% of 6.
[b] With addition of 10 g of oil or 5 g of oil and 10 g sugar or 20 g sugar.
[c] With addition of 5 g of oil or 10 g sugar.

Nutritional value of the components

Table 2.8, which shows the composition of potatoes cooked by different methods, indicates that potatoes baked in their skins or roasted or fried in fat, can make much greater contributions to intakes of energy and protein than when boiled. This is wholly (in the case of baking) or partly (in the case of roasting or frying) due to a concentration of nutrients through water loss during cooking. Such figures do not show nutrient losses that may occur in the process and these will be discussed fully in Chapter 4. Such methods of cooking are expensive and more extensively used in the developed world, where excessive amounts of energy are often consumed, than in the developing world, where extra energy is frequently required.

Table 2.8. *Compositon of potatoes cooked by different methods (per 100 g)*

Cooking method	Energy (kJ)	Energy (kcal)	Moisture (%)	Crude protein (g)	Fat (g)	Total carbo-hydrate (g)	Dietary fibre (Crude fibre) (g)
Uncooked	335	80	78.0	2.1	0.1	18.5	1.7[a] (0.5)
Boiled in skin[b] (flesh only)	318	76	79.8	2.1	0.1	18.5	(0.5)
Boiled, peeled[d]	301	72	81.4	1.7	0.1	16.8	1.6[a] (0.6)
Baked in skin[f] (flesh only)	414	99	73.3	2.5	0.1	22.9	1.9[a] (1.2)
Mashed (with milk and margarine)[f]	444	106	78.4	1.8	4.7	15.2	(0.7)
Roasted (in shallow fat, flesh only)[a]	657	157	64.3	2.8	4.8	27.3	2.7[a]
French fried (chips)[f]	1165	264	45.9	4.1	12.1	36.7	3.3[a] (1.0)
Chips (crisps)[f]	2305	551	2.3	5.8	37.9	49.7	11.9[e] (1.6)

[a] Finglas & Faulks (1984).
[b] Watt & Merrill (1975).
[c] Calculated as by Paul & Southgate (1978).
[d] Average of figures from Paul & Southgate (1978), Watt & Merrill (1975), Wu Leung *et al.* (1978).
[e] Paul & Southgate (1978).
[f] Average of figures from Paul & Southgate (1978), Watt & Merrill (1975).
[g] Estimated values from Paul & Southgate (1978).

Energy and protein

Table 2.9 shows the composition of various processed forms of potatoes including those produced by Western industrial methods (A) and those produced by traditional methods in the highlands of Peru (B). The energy contributions of such products, especially the traditionally produced, dried ones, will depend upon the form in which they are cooked, i.e. the concentration of potato in the final food as eaten. It can be seen that the protein content of some processed products is considerably reduced. A dried form of potato, known locally in Andean countries as *chuño blanco*, has an average crude protein content of only 1.9% (DWB), compared to the 8.4% of dried potato, as shown in Table 2.3. This reduction could be important in Peruvian communities relying on supplies of *chuño blanco*

Ash (g)	Ca (mg)	P (mg)	Fe (mg)	Thiamin (mg)	Niacin (mg)	Ribo-flavin (mg)	Pyri-doxine (mg)	Total folic acid (μg)	Ascorbic acid (mg)
1.0	9	50	0.8	0.10	1.5	0.04	0.25	14	20
				(0.2)[a]	(0.6)[a]	(0.02)[a]		(35)[a]	
0.9	7	53	0.6	0.09	1.5	0.03	—	—	12–16[c]
0.7	6	38	0.5	0.08	1.2	0.03	0.18	10	4–14[e]
				(0.2)[a]	(0.5)[a]	(0.01)[a]		(30)[a]	
1.2	10	60	0.8	0.10	1.8	0.04	0.18	10	12–16[c]
				(0.2)[a]	(0.6)[a]	(0.01)[a]		(25)[a]	
1.5	18	40	0.4	0.08	1.1	0.04	0.18	(10)[g]	4–12[e]
—	10	53	0.7	0.10	1.9	0.04	0.18	7[e]	5–16[e]
				(0.2)[a]	(0.6)[a]	(0.02)[a]		(35)[a]	
1.8	15	92	1.1	0.12	2.6	0.06	0.18	(10)[g]	5–16[e]
				(0.2)[a]	(0.6)[a]	(0.02)[a]		(35)[a]	
3.1	39	135	2.0	0.20	5.5	0.07	0.89	20[e]	17[e]

during parts of the year when supplies of fresh potatoes are limited. Protein content is not reduced to the same extent during processing of *chuño negro*, and Gursky (1969) has recorded that *chuño negro* supplied 5.9% of the protein intake in the three Peruvian highland communities which he studied. The preparation of *chuño* is described briefly in Chapter 4.

Table 2.9. *Composition of processed forms of potatoes (per 100 g)*

Form of potato	Energy (kJ)	Energy (kcal)	Moisture (%)	Crude protein (g)	Fat (g)	Total carbohydrate (g)	Dietary fibre (Crude fibre) (g)
A. *Industrial methods*							
Canned[a] (drained solids)	222	53	84.2	1.2	0.1	12.6	2.5[a]
Instant powder[a]	1331	318	7.2	9.1	0.8	73.2	16.5[a]
Instant powder[a] (reconstituted)	293	70	79.4	2.0	0.2	16.1	3.6[a]
Frozen (diced)[d]	305	73	81.0	1.2	Trace	17.4	(0.4)
B. *Traditional methods*							
Frozen (by traditional Peruvian method)[e]	753	180	54.5	1.8	0.6	42.1	(2.0)
Papa seca[e,f]	1347	322	14.8	8.2	0.7	72.6	(1.8)
Chuño blanco (white *chuño*)[e,g]	1351	323	18.1	1.9	0.5	77.7	(2.1)
Chuño negro (black *chuño*)[e,h]	1393	333	14.1	4.0	0.2	79.4	(1.9)

[a] Paul & Southgate (1978).
[b] Estimated values from Paul & Southgate (1978).
[c] Some brands are fortified and may contain up to 10 times this value.
[d] Watt & Merrill (1975).
[e] Collazos *et al.* (1974).
[f] Potatoes boiled, peeled and sun-dried.
[g] Potatoes frozen outdoors, trampled, soaked in running water for 1–3 weeks, sun-dried.
[h] Potatoes frozen outdoors, trampled, sun-dried.

Ash (g)	Ca (mg)	P (mg)	Fe (mg)	Thiamin (mg)	Niacin (mg)	Ribo-flavin (mg)	Pyri-doxine (mg)	Total folic acid (μg)	Ascorbic acid (mg)
	11	31	0.7	0.02	1.0	0.03	0.16	11	17
	89	220	2.4	0.04	5.6	0.14	(0.82)[b]	24	12[c]
	20	48	0.5	0.01	1.7	0.03	(0.18)[b]	5	3[c]
0.4	10	30	0.7	0.07	0.6	0.01	—	—	9
1.0	58	54	2.8	0.07	1.6	0.20	—	—	1
3.5	47	200	4.5	0.19	5.0	0.09	—	—	3
1.8	92	54	3.3	0.03	3.8	0.04	—	—	1
2.3	44	203	0.9	0.13	3.4	0.17	—	—	2

Dietary fibre

In recent years there has been increasing interest in dietary fibre, as a result of suggestions that it gives protection against diverticulosis, cardiovascular disease, colonic cancer and diabetes: this is the subject of continuing intensive research.

Trowell *et al.* (1976) stated that dietary fibre is 'the plant polysaccharides and lignin which are resistant to hydrolysis by the digestive enzymes of man'; this is the definition most widely accepted at present. Dietary fibre is not, however, a precise term and opinions vary on its exact composition. Methods of determining dietary fibre are continually being modified and improved (Hellendoorn *et al.*, 1975; Jeltema & Zabik, 1980; Schweizer & Wursch, 1981). Recent methods show that the quantity of dietary fibre in foods, including potatoes, is higher than that of crude fibre. However, this difference is due to the variation in methodology used in the determinations. Dietary fibre analyses utilize physiologically active enzymes to break down non-fibre components, whereas earlier chemical determinations of crude fibre used acid or alkali. Crude fibre analyses have largely been abandoned as they measure only a small and variable fraction of the dietary fibre. However, crude fibre contents of foods are still generally given in food composition tables. Where data are available, contents of dietary fibre are given in the tables in this chapter, although few determinations have been carried out and published results vary considerably. For example, raw potato dietary fibre content apparently ranges between about 1 g and 2 g/100 g fresh weight (Paul & Southgate, 1978; Finglas & Faulks, 1984a; Wills *et al.*, 1984; Jones *et al.*, 1985). In addition, it should be noted that part of the dietary fibre may be starch that is resistant to hydrolysis by the enzymes used to remove starch prior to dietary fibre determination. This 'resistant starch' may be produced by subjecting foods to heat and/or dehydration, which confers a more ordered structure on the starch molecules and renders them less susceptible to enzymic digestion. Jones *et al.* (1985) found that there was little 'resistant starch' in raw potato, but that it formed 20% to 50% by weight (determined separately) of the total dietary fibre of cooked potato. Variations in dietary fibre as affected by rate of starch digestibility *in vitro* were also determined in raw, baked and chipped (crisps) potatoes by Dreher *et al.* (1983). However, it is as yet unknown whether this resistant starch is digested in the human intestine. If it is not, then it should be considered as part of the dietary fibre since it may, like other types of fibre, affect health through the medium of the colon.

It has been shown that various types of dietary fibre have differing physiological effects. For example, insoluble cereal fibre affects transit time and faecal weight, whereas soluble gel-forming fibre has been shown to reduce serum cholesterol levels and blood glucose and insulin response to meals containing carbohydrates (Sandberg, 1982). A chemical characterization of the dietary fibres in various foods, including whole potatoes (Theander, 1983), revealed that cereal brans have the highest lignin values, are rich in arabinoxylans and cellulose, but are low in uronic acids, whereas various vegetables, including potatoes, have higher cellulose contents and more pectic and pectin-associated substances than the brans. Other recent research on this topic has been initiated by Englyst *et al.* (1982), Reistad (1983) and Jones *et al.* (1985). The significance of the findings in terms of the nutritional properties of different dietary fibre sources, however, has yet to be determined. For this reason, fibre content of potatoes and other foods will be compared here only quantitatively.

By comparison with other raw items (Tables 2.2 and 2.3), the fresh potato has a dietary fibre content similar to that of sweet potato but somewhat lower than that of other roots and tubers and, much lower than that of most cereals and dry *Phaseolus* beans, although, on a dry basis, potatoes and cereals are similar in this respect. Dietary fibre determinations have been conducted largely on raw rather than on cooked foods and information on levels was found for only five of the ten cooked items in Table 2.4. Boiled potato flesh has a dietary fibre content similar to that of cooked white rice and a much lower content than that of boiled green plantains or of boiled *Phaseolus* beans. Potatoes cooked as french fries or chips are a more concentrated source of fibre (Table 2.8). It can be calculated that 100 g of boiled potato supply 1.0, 0.7, and 0.5 times the fibre that can be found in a 35 g 'medium' slice of white, brown or wholemeal bread, respectively; a 25 g packet of chips supplies 1.9, 1.4 and 1.0 times the respective bread fibre contents.

There is no recommended daily allowance (RDA) for dietary fibre at present. It has been suggested (Brodribb, 1983) that about 40 g/day should be consumed to maintain correct colonic function. Recently an *ad hoc* working party of the NACNE (National Advisory Committee on Nutrition Education, 1983) recommended an increase in British dietary fibre intake to 30 g per person per day. When potatoes are consumed in quantity on a regular basis, they make a significant contribution to dietary fibre intake. At present, for example, fresh potatoes contribute 15% of the dietary fibre intake in British households and rank as a primary source (Finglas & Faulks, 1985).

Consumption of the whole tuber, rather than the flesh alone, may increase dietary fibre intake. Jones *et al.* (1985) found a higher dietary fibre content in unpeeled than in peeled raw or boiled potatoes (see

Table 2.4). There have been investigations into whether peel, as a by-product of the potato-processing industry in developed countries, could be incorporated into bread to increase fibre in the diet generally (Toma *et al.*, 1979; Orr *et al.*, 1982).

Vitamins

Potato is an excellent source of some of the water-soluble vitamins. Average values of the contents of these vitamins in the raw potato are shown in Tables 2.2, 2.10 and 2.11.

Factors affecting contents

Values can vary considerably, as the ranges, determined by different authors, shown in Table 2.10 demonstrate, but relatively little work has been carried out to determine the sources of variation. Different methods of analysis can lead to varying results: Finglas & Faulks (1984*a*, 1985) attributed differences in their determined values for thiamin, niacin, riboflavin and total folate from those previously reported for the potato in food composition tables to analytical methods that were more reproducible than those used earlier.

The thiamin content of potatoes depends upon variety (Swaminathan & Pushkarnath, 1962; Leichsenring *et al.*, 1951) and location of growth (Leichsenring *et al.*, 1951). Tubers from loamy soil contained more thiamin than tubers from sandy soil, and thiamin content is greatly increased by nitrogen fertilization (Augustin, 1975).

A large difference in riboflavin content, but little difference in niacin content, has been found amongst varieties (Swaminathan & Pushkarnath, 1962). Riboflavin was unaffected by soil type, and little affected by nitrogen fertilization when potatoes were grown on sandy soils, but was increased somewhat by nitrogen fertilization on loamy soils. Niacin was unaffected by soil type, but was increased by nitrogen fertilization on both sandy and loamy soils (Augustin, 1975).

More information is available on factors determining variation in potato ascorbic acid (vitamin C) content. References to early work on these factors are given by Leichsenring *et al.* (1951) and Burton (1966). A series of experiments in the 1940s (Leichsenring *et al.*, 1951) showed that the reduced ascorbic acid content of potatoes varies with variety, locality (the amount of the difference between localities being dependent upon the variety under consideration), crop year and maturity at time of harvest (values were highest when plants were at their maximum vigour and declined thereafter as vines began to die off).

Recent studies have confirmed these results and showed an additional

Table 2.10. *Ranges in contents of various water-soluble vitamins in potatoes (mg/100 g FWB)*[a]

Reference	Thiamin	Riboflavin	Niacin	Ascorbic acid	Folic acid	Pyridoxine
Swaminathan & Pushkarnath (1962)	0.031–0.078 (0.051)[b]	0.012–0.076 (0.033)	0.9–3.2 (1.6)	13.1–26.6 (19.6)	—	—
Page & Hanning (1963)	—	—	1.03–2.08 (1.54)	—	—	0.13–0.42 (0.26)
Szkilladziowa et al. (1977)	0.06–0.10 (0.087)	0.031–0.063 (0.046)	0.85–1.87 (1.46)	—	0.036–0.053 (0.041)	0.225–0.370 (0.300)
Augustin et al. (1978)	0.06–0.12 (0.09)	0.020–0.092 (0.038)	0.7–2.6 (1.5)	7.8–36.1 (14.1)	0.006–0.020 (0.013)	0.126–0.400 (0.257)
Finglas & Faulks (1984a)	0.1–0.2 (0.2)	0.02–0.03 (0.02)	0.5–0.8 (0.6)	7–19	0.025–0.050 (0.035)	—

[a] Where necessary, figures have been re-expressed from the original values, in mg/100 g (FWB).
[b] Values in parentheses are means of ranges given above.

dependency on soil type and nitrogen and phosphorus fertilization. Large variations were found between 12 varieties for each of three locations (Augustin, 1975), the highest reduced ascorbic acid content being almost twice that of the variety with the lowest content.

Date of harvest influenced reduced ascorbic acid levels in 'Russet Burbank' tubers (Augustin *et al.*, 1975). Early harvest (103 days after planting) resulted in high ascorbic acid values, whereas increasing delays in time of harvest, up to 166 days after planting, resulted in gradual decreases in values. 'Normal' harvest date was considered to be 150 days after planting. The same authors showed that 'Russet Burbank' tubers grown on sandy soils contained greater quantities of ascorbic acid than those grown on loamy soils.

Clarification is needed on effects of nitrogen fertilization on ascorbic acid levels, as reports in the literature are conflicting. Two early studies quoted by Burton (1966) found no significant effect of manuring on content of ascorbic acid. However, nitrogen fertilization has been shown generally to decrease reduced ascorbic acid values in 'Russet Burbank'

Table 2.11. *Vitamin and mineral contents of potatoes and of some internationally important vegetables (per 100 g raw)*[a]

Vegetable	Carotene (μg)	Thiamin (mg)	Riboflavin (mg)	Niacin (mg)	Potential niacin (Tryptophan ÷ 60) (mg)	Ascorbic acid (mg)
Potato (freshly harvested)	0–trace	0.11[b]	0.04[b]	1.2[b]	0.5	30[b]
Carrot	12 000	0.06	0.05	0.6	0.1	6
Onion	0	0.03	0.05	0.2	0.2	10
Tomato	600	0.06	0.04	0.7	0.1	20
Pepper (green)	200	Trace	0.03	0.7	0.2	100
Pumpkin	1500	0.04	0.04	0.4	0.1	5
Okra	90	0.10	0.10	1.0	0.3	25
Green beans (runner)	400	0.05	0.10	0.9	0.4	20
Cauliflower	30	0.10	0.10	0.6	0.5	60

[a] Paul & Southgate (1978).
[b] See also values given in Table 2.10.
[c] Estimated values.

tubers (Augustin *et al.*, 1975). Response of ascorbic acid to nitrogen fertilization can depend upon variety. Two out of three varieties analysed showed decreased reduced ascorbic acid levels with increasing nitrogen fertilization, whilst the third variety was unaffected (Augustin, 1975). In contrast, Mondy *et al.* (1979) found that reduced ascorbic acid increased significantly with increasing nitrogen levels of ammonium nitrate at 100 to 250 lb/acre (114 to 284 kg/ha).

The influence of phosphorus may depend upon the level applied. Application of 112 kg monoammonium phosphate/ha produced a significant increase in reduced ascorbic acid (Klein *et al.*, 1980), whereas lower levels of phosphorus had no consistent effect (Teich & Menzies, 1964).

McCay *et al.* (1975) cited various Russian studies which have reported increases in ascorbic acid concentration with applications of trace elements, such as molybdenum, boron, manganese and copper, to the soil, and another study which showed increased ascorbic acid when potato plants were sprayed with various desiccants.

Pyri-doxine (mg)	Folic acid		Panto-thenic acid (mg)	Biotin (μg)	Ca (mg)	P (mg)	Fe (mg)
	Free (μg)	Total (μg)					
0.25	10	14^b	0.30	0.1	8	40	0.5
0.15	12	15	0.25	0.6	48	21	0.6
0.10	15	16	0.14	0.9	31	30	0.3
0.11	15	28	0.33	1.5	13	21	0.4
0.17	5	11	0.23	—	9	25	0.4
0.06	(13)^c	(13)^c	0.40	(0.4)^c	39	19	0.4
0.08	25	100	0.26	—	70	60	1.0
0.07	57	60	0.05	0.7	27	47	0.8
0.20	30	39	0.60	1.5	21	45	0.5

Nutritional value of the components

Contribution to the diet

Potato is a substantial source of ascorbic acid, thiamin, niacin and pyridoxine and its derivatives (vitamin B_6 group) and also contains folic acid, pantothenic acid (vitamin B_5) and riboflavin. Comparisons may be made either with staple foods (Tables 2.2, 2.3 and 2.4) or with vegetables (Table 2.11), because although potatoes are usually regarded as a staple food, in many parts of the developing world they are eaten in small quantities as a vegetable. It should be noted that recent analyses of British potatoes (Finglas & Faulks, 1984*a*), given in Table 2.10, show higher values for thiamin and folate and lower values for niacin and riboflavin than those given in previous food composition tables. Thiamin and riboflavin (*ibid.*, 1984*b*) and niacin (*ibid.*, 1984*a*) were determined by high performance liquid chromatography, and folic acid by an improved microbiological method (*ibid.*, 1984*a*). It is possible that vitamin values for the other staples and vegetables may also have to be revised in the light of new analytical techniques.

At present, potato may be seen from the tables to compare favourably on a raw basis with all the listed staples and vegetables in terms of thiamin, riboflavin and niacin, and with most of the vegetables in pyridoxine and pantothenic acid contents. It has a much lower biotin content than the other vegetables, but it may be a comparatively richer source of folic acid than was previously thought. Fresh potatoes may contain 30 mg or more of ascorbic acid per 100 g when newly harvested, with an average value of 20 mg/100 g, although values decline when potatoes are stored, cooked or processed. This aspect is discussed more fully in Chapter 4. Sweet potatoes, cassava and plantain are comparable to potatoes in terms of ascorbic acid content (Tables 2.2, 2.3 and 2.4), but yam and cocoyam have much lower quantities and most cereals (with the exception of some forms of maize) and beans are totally lacking in this vitamin unless sprouted. From Table 2.12, it can be seen that potato has higher quantities of ascorbic acid than do carrots, onions and pumpkin, quantities similar to those of runner beans, okra and tomato, but only half and about one-third as much as in cauliflower and green peppers, respectively. The potato's role as a source of some of the B-vitamins and of ascorbic acid, when eaten either as a vegetable or as a staple, has been greatly underestimated by potato consumers. Burton (1974) mentions a survey among British housewives in which only 2% of those interviewed regarded the potato as a source of vitamins or protein. In terms of human requirements, Table 2.12 shows the substantial contribution made by only 100 g potato, boiled in its skin, to the RDAs for thiamin, niacin, folic acid, pantothenic and ascorbic acids. In the light of recent analyses,

Vitamins

already mentioned, potatoes may provide much bigger percentages of the thiamin and folic acid RDAs than are shown here. They may, however, contribute rather less niacin. Furthermore, only 23% of the total niacin in cooked (baked) potatoes was found, by rat bioassay, to be in an available form (Carter & Carpenter, 1980). By comparison, 30% to 40% of the niacin in mature cereal grains and peanuts, and 100% of the niacin in beans, were available. Niacin availability in boiled and other cooked forms of potato deserves further investigation.

With an average figure of 16 mg ascorbic acid/100 g, 100 g of freshly harvested potatoes, boiled in their skins, furnish 80% of a child's and 50% of an adult's RDA (Table 2.12). Such percentages fall when ascorbate levels are depleted following storage or cooking after peeling. The vital contribution potato can make to dietary ascorbic acid intake emphasizes the need to conserve ascorbic acid levels by careful domestic or large-scale cooking methods.

Table 2.12. *Percentages of recommended daily allowances of major nutrients provided by 100 g of cooked potato (boiled in skin)*[a]

Age group	Thiamin	Niacin	Folic acid[b]	Pyridoxine[c]	Ascorbic acid	Fe
Children						
1–3 years	18	17	12	25	80	6–12
4–6 years	13	12	12	18	80	6–12
7–9 years	10	10	12	14	80	6–12
Male adolescents						
10–12	9	9	12	13	80	6–12
13–15 years	8	8	6	13	53	3–7
16–19 years	8	7	6	12	53	7–12
Female adolescents						
10–12 years	10	10	12	13	80	6–12
13–15 years	9	9	6	13	53	3–5
16–19 years	10	10	6	12	53	2–4
Adult man (moderately active)	8	8	6	10	53	7–12
Adult woman (moderately active)	10	10	6	12	53	2–4

[a] Unless otherwise indicated, calculated from the figures for cooked potato given in Table 2.4 as percentages of RDAs given by Passmore *et al.* (1974).
[b] Calculated from figure for folic acid given in Augustin *et al.* (1978).
[c] Calculated from figure for pyridoxine given by Augustin *et al.* (1978) as % USRDA.

The importance of potato as a vitamin source, particularly of ascorbic acid, has been demonstrated clearly in several developed countries. The range of total ascorbic acid levels over a 12-month period found in several varieties of potatoes grown in Australia was estimated to provide 50% to 160% of the Australian daily allowance at the average daily consumption rate of 150 g per person (Wills et al., 1984). In Britain, potato contributed 19.4%, 8.7% and 10.6% of total household intakes of ascorbic acid, thiamin and niacin, respectively, in 1983 (National Food Survey Committee, 1983) and was the primary single food source of ascorbate. It has also been estimated to have provided at least 12% of British folate intake in 1984 (Finglas & Faulks, 1985) and 28% and 11% of the British pyridoxine and pantothenic acid intakes in 1979 (Finglas & Faulks, 1985; more up-to-date figures are not yet calculated). The proportion of ascorbic acid intake from potato in various parts of Europe was estimated to range from 10% in the south to 50% to 60% in the north-east and far north (Burton, 1974). Potatoes contribute as much ascorbic acid (20%) to the United States food supply as citrus fruits (18%) (McCay et al., 1975), although the latter have greater concentrations of ascorbic acid. The valuable contribution that early varieties (new potatoes) can make in boosting dietary ascorbic acid intake when levels in maincrop (old) potatoes have fallen to low levels due to many months of storage, has been shown for British potatoes (Finglas & Faulks, 1984a). In one season, the mean total ascorbic acid content of old potatoes stored for 9 months and of new potatoes were found to be 6.7 mg and 18.6 mg per 100 g, respectively.

Table 2.9 demonstrates that processing of potatoes can reduce or almost totally destroy ascorbic acid depending upon the method of processing. (This is discussed fully in Chapter 4.) Reduction of ascorbic acid to low levels in potatoes processed by traditional methods in Peru may be of importance in limiting ascorbic acid supplies when fresh potatoes are unavailable at certain times of the year.

Data on the levels of ascorbic acid, thiamin, riboflavin, niacin, folic acid and pyridoxine in fresh and cooked US commercial varieties of potatoes have been published by Augustin and coworkers (1975; Augustin et al., 1978). At present there is no information on the contributions, potential or actual, by potato to vitamin supplies in developing countries. Where diets are largely based on cereal staples and dry legumes lacking in ascorbic acid, potatoes might be used as a supplement.,

Since potatoes have a low lipid content, the fat-soluble vitamins are absent or are present only in trace amounts. The potato would not, therefore, be of use where vitamin A deficiency is a public health problem, the only rich staple food sources of this vitamin being sweet potato, yellow varieties of maize, and plantains (Tables 2.2 and 2.3). All vegetables listed in Table 2.11, with the exception of potato, contain some vitamin A, and carrots are a rich source. A mixed diet containing potato should therefore include a source of vitamin A.

Minerals and trace elements

The ash content of potatoes is about 1% (FWB) and contains some important minerals and trace elements essential to various human body structures and functions. Determinations of the major minerals and trace elements in raw potatoes have been carried out by various researchers (True *et al.*, 1978; Wolnik *et al.*, 1983; Finglas and Faulks, 1984a; Wills *et al.*, 1984); average contents and ranges found are shown in Table 2.13.

Table 2.13. *Minerals and trace elements in raw tubers (mg/100 g FWB)*

	Mean (unpeeled)[a]	Mean (peeled)[b]	Ranges[c]
Minerals			
Calcium	6.5	5.5	1.7–18
Magnesium	20.9	18.6	10–29
Phosphorus	47.9	44.0	27–89
Potassium	564.0	376.0	204.9–900.5
Sodium	7.7	6.6	2–66
Trace elements			
Aluminium	0.610	—	0.301–1.511
Boron	0.136	—	0.081–0.168
Chromium	0.023[d]	—	—
Cobalt	0.065[e]	—	—
Copper	0.193	0.088	0.014–0.327
Fluorine	—	0.11[f]	0.02–0.38
Iodine	0.019	—	0.011–0.035
Iron	0.740	0.403	0.13–2.311
Manganese	0.253	0.14[g]	0.072–0.699
Molybdenum	0.091	0.0036[g]	<0.011–0.186
Nickel	—	—	0.008–0.037
Selenium	0.006	0.0003[g]	<0.0002–0.029
Zinc	0.410	0.280	0.11–0.70

[a] True *et al.* (1978), except where indicated.
[b] Average of Wolnik *et al.* (1983) and Finglas & Faulks (1984a), except where indicated.
[c] Maximum ranges found by researchers.
[d] Zawadzka (1978); whole, unpeeled?
[e] Lampitt & Goldenberg (1940), on a dry weight basis.
[f] Walters *et al.* (1983).
[g] Wolnik *et al.* (1983).

Factors affecting contents

Although wide ranges have been encountered in contents of many minerals and trace elements in potatoes (Table 2.13), there is little information about what determines these levels. True *et al.* (1978), in a brief review, cite work attributing variations in mineral content to such factors as soil type, location of growth and application of phosphorus.

In addition, varietal differences have been noted in the contents of calcium, phosphorus and iron in 13 varieties grown in the same location (Leichsenring *et al.*, 1951). A more recent study also found varietal variations in iron and in manganese, but not in zinc and copper levels (Kubisk *et al.*, 1978).

Contents of calcium, phosphorus and iron were also influenced by location and this was more important than variety in the case of calcium (Leichsenring *et al.*, 1951). Wide ranges in some mineral elements (namely calcium, phosphorus, sodium, potassium, selenium and aluminium) observed in nine varieties grown at five locations were attributed to location of growth rather than to variety (True *et al.*, 1978). However, it was noted that these differences could be due to various factors, including mineral content of the soil, cultivation practices and sampling procedures.

Nitrogen fertilization had little effect on magnesium, calcium, potassium, sodium and phosphorus levels in 'Russet Burbank' tubers (Augustin, 1975). However, iron content was increased somewhat. The same author also reported that tubers grown on sandy soils had lower quantities of magnesium than those grown on loamy soils, but it was concluded that the higher iron content in tubers from sandy soil was a result of the iron content of the soil rather than of the soil type.

Contribution to the diet

In potato, *iron* content is about the same as that in other roots and tubers (Table 2.2) and some vegetables (Table 2.12) and is comparable on a dry weight basis to that in some of the cereals (Table 2.3). It is higher than iron levels in white rice, on either a dry or cooked basis. Though not an outstanding source of iron, 100 g of cooked potatoes can supply between 6% and 12% of daily iron requirements for children or adult men (Table 2.12). The percentage contributions are lower in women of child-bearing age, due to the greater demand for iron caused by menstruation, pregnancy and lactation. Such figures, however, do not take into account availability of the iron. In foods, iron is either in haem or non-haem form. The National Research Council (1980) treats the iron in vegetable products, including potatoes, as non-haem, the availability of which may be enhanced by the presence of ascorbic acid ingested at the same time as the iron source. Potato ascorbic acid may contribute towards the level needed to influence iron absorption from a meal. A positive correlation was found between ascorbic acid content of potatoes and the amount of iron solubilized from potatoes by gastric juice *in vitro* (Fairweather-Tait, 1983). Only half as much iron (15%) was available from potato (depleted in ascorbic acid content by previous cooking, drying and grinding) as from well-absorbed $FeSO_4$ (32%) in a rat *ad libitum* feeding experiment. However, a much higher proportion of the iron from potato was solubilized *in vitro* than there was from other vegetable foods such as kidney beans, wheat flour and bread. It has been suggested that iron solubilization is the first step in determining iron availability from a food or meal. Hence potato appears to have a moderate iron availability superior to that of other vegetable foods (Fairweather-Tait, 1983).

In Britain, potatoes have been shown to supply 6% of the total household dietary iron intake, ranking third of all individual foods as a dietary source (National Food Survey Committee, 1983). True *et al.* (1978) found that 150 g of potato could supply 2.3% to 19.3% of the USRDA for iron. Although, in a later study, these authors indicated that fresh potatoes on a 150 g serving basis did not contribute any iron to the USRDA (True *et al.*, 1979). This difference is undoubtedly due to variation in iron content found by the authors amongst varieties which were grown in several different locations. Also it should be noted that the USRDA for iron is very high. With the more realistic RDA given by FAO/WHO, potato makes a significant contribution.

Potatoes are a good source of *phosphorus*, being similar, in this respect, to roots and tubers and most cereals on a cooked basis. Tortillas, bread and boiled *P. vulgaris* beans are richer sources of phosphorus than are potatoes. However, 100 g of boiled potato supplies 7% of the USRDA for phosphorus for both children and adults.

It is important to note that a relatively small percentage of the total phosphorus in potatoes occurs in the form of phytic acid (a hexaphosphate derivative of inositol). Phytic acid is insoluble and cannot be absorbed in the human intestines. It also binds calcium, iron and zinc in the form of phytates, thus rendering them unavailable for absorption into the body. Approximately 25% of the total phosphorus was found in phytic acid in seven commercial North American potato varieties (Quick & Li, 1976). Lampitt & Goldenberg (1940) quote two sources which found that at least 80% of the phosphorus in potatoes was non-phytic. A mean of only 8.3% of the total phosphorus was found in phytic acid amongst 23 samples of potatoes grown in India (Swaminathan & Pushkarnath, 1962). In contrast, other plant foods contain much higher levels of phytic acid. For example, Akroyd & Doughty (1970) noted that the percentage of phytic acid phosphorus in total phosphorus of high and low extraction wheats was 75% and 68%, respectively. In samples of field beans (*Vicia faba*), 40% to 60% of the total phosphorus was in the form of phytate phosphorus (Griffiths & Thomas, 1981). The lower phytic acid content of potatoes therefore may be advantageous in allowing greater availability of the phosphorus, calcium, iron and zinc which may be present in a meal which includes potatoes. This is especially important in the case of *calcium*. Potato is a poor source of calcium, a characteristic it shares on a cooked basis with the other cooked staples (Table 2.4), with the exception of lime-treated tortillas and *P. vulgaris* beans and other legumes. Apart from okra, none of the vegetables listed in Table 2.11 is a particularly good source of calcium.

The concentration of *potassium* in potatoes is high (Table 2.13), and for this reason potatoes are frequently omitted from the diets of patients with renal failure (McCay *et al.*, 1975). Conversely, sodium content is low, and potatoes cooked unpeeled do not absorb any sodium chloride from the cooking water. Potatoes can therefore be used in diets designed to restrict sodium intake, for example, in patients with high blood pressure, where a high potassium:sodium ratio may be of additional benefit.

Magnesium is another important dietary mineral. For raw potatoes, 150 g were found to provide between 6% and 10% of the USRDA for magnesium (True *et al.*, 1979) and this range is likely to be the same for cooked potatoes as there is almost 100% retention of the mineral in potatoes boiled in their skins (True *et al.*, 1979).

In the United States, RDAs have been established for only two of the trace elements found in potatoes: *zinc* and *iodine*. Comparing figures for levels of zinc and iodine in potato shown in Table 2.13 with the USRDA values, 100 g of potato should provide 13% of adult, and up to 30% of child requirements of iodine and 2% and 4% of adult and child requirements, respectively, of zinc. The National Research Council (1980) has noted that biological availability of zinc is generally lower in vegetables in foods derived from animals, and depends in part upon the contents of phytic acid and dietary fibre in the source. Unlike whole grain products, potato does not contain excessive quantities of dietary fibre and its phytic acid content is relatively low, so zinc availability should be high. Maga (1983) found that 97% of zinc was available in a potato-based diet for rats, with a phytic acid content of 0.23 mg/100 g, whereas only 23% was available in a corn-based diet with a phytic acid content of 9.93 mg/100 g.

Other, less extensively investigated trace elements reported to have beneficial effects in humans include copper, chromium, manganese, selenium and molybdenum (Passmore *et al.*, 1974; National Research Council, 1980); for these the National Research Council (1980) provides a table of ranges for recommended intakes. If this table is compared with Table 2.13, it may be calculated that 100 g of potatoes can supply at least part of the daily requirement for copper, manganese, molybdenum and chromium. True *et al.* (1979) recommended that potatoes be considered as supplying 8% of the USRDA for copper per 150-g serving. Wenlock *et al.* (1979) determined that the manganese content of potatoes ranged from 0.7 to 1.9 mg/kg (p.p.m.) and that it made a partial contribution to the 15% of daily intake provided by all fruits and vegetables in the British diet. According to a table of contributions (per person per day) made by foods to selected minor nutrients in British household food in 1976 (Spring *et al.*, 1979), potatoes provided 10% of the magnesium, 11% of the copper and 3% of the zinc of British diets. A more recent analysis revised these contributions to 8.4% and 4.3% for copper and zinc, respectively (Finglas & Faulks, 1984*a*, 1985). Lastly, potatoes are not a good source of selenium (Thorn *et al.*, 1978), containing less than 0.01 mg/kg.

Further trace elements for which no deficiencies have yet been found in man, nor have any requirements been established, but in which some interest has been shown, include cobalt and nickel (Table 2.13) and vanadium. The content of vanadium was less than 1 ng/g (Myron *et al.*, 1977).

Potato is not an outstanding source of fluoride (Table 2.13), and in this respect is similar to most other foods (Walters *et al.*, 1983).

Summary

Potato is usually considered to be a rich source of energy and to provide negligible amounts of protein, vitamins and minerals. It has been shown here that, by itself, it is a poor source of energy unless fried, but that it contains good-quality protein, many of the water-soluble vitamins, and some minerals and trace elements. If the body's energy requirements are met, potato can also provide a significant amount of dietary protein. However, its low energy density means that, if potato is consumed alone, a considerable bulk must be eaten just to satisfy energy requirements. This is particularly disadvantageous in the case of small children, whose capacity is limited.

Although protein content compares unfavourably on a raw basis with cereals such as rice, on a cooked basis it is similar to that of cooked rice. Potato protein is moderately limiting in the sulphur-containing amino acids, but contains substantially more lysine than the cereal staples and can therefore be a useful supplement. Potato may be more attractive as a source of high-quality protein rather than energy, in those parts of the tropical developing world where it must compete, as a food, with other well-established roots and tubers. As little as 100 g can supply a significant percentage of the daily protein requirement from childhood onwards (Figure 2.4).

When eaten in quantity on a regular basis, potatoes can make a useful contribution to dietary fibre intakes, especially if both the skin and the flesh are consumed.

The tuber is a good source of water-soluble vitamins, including some of the B group vitamins and ascorbic acid (vitamin C). When boiled in its skin, 100 g can make a valuable contribution to the RDAs of thiamin,

Figure 2.4. A young Peruvian snacks on a potato boiled in its skin.

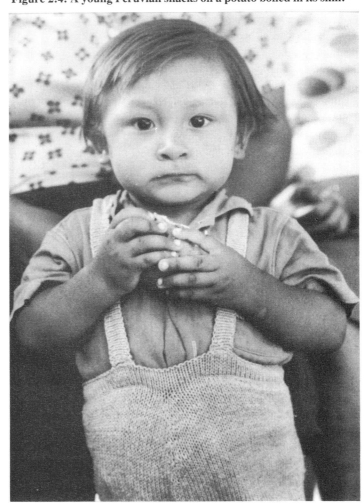

Summary

niacin, pyridoxine, folic acid and particularly ascorbic acid. The availability of niacin in baked potatoes is low and therefore it should be investigated in other forms of cooked potato. In contrast to cereals and legumes, which lack ascorbic acid, potato has been shown to contribute substantially to the daily intake of this vitamin in various developed countries. As a vegetable, it is comparable to many other vegetables in terms of contents of ascorbic acid, thiamin, niacin, pyridoxine and pantothenic acid.

Amongst the many minerals present in the tuber, iron, magnesium and phosphorus contribute significantly to daily requirements. In potato, the percentage of total phytic acid is much lower than in the cereals, and thus has little effect on the availability of minerals such as iron, zinc and calcium, which is present at only a low level. Potatoes can supply a high percentage of iodine requirements (although this probably depends on the iodine content of the soil in which the potatoes are grown), smaller percentages of zinc requirements, and can make at least some contribution to supplies of copper, manganese, molybdenum and chromium.

Potatoes are rarely eaten alone. Their importance in this respect depends upon the area of the world and the particular meal under consideration. In some parts of the Peruvian sierra, a meal may consist entirely of potatoes. In India they are often nothing more than a small vegetable accompanying the major ingredients. Potatoes are also eaten as a snack food in some parts of the world (Figure 2.5). In main meals or

Figure 2.5. A tea-time snack of potatoes boiled in their skins in eastern Java, Indonesia.

snacks they can provide at least part of the body's requirements for many essential nutrients.

References

Augustin, J. (1975). Variations in the nutritional composition of fresh potatoes. *J. Food Sci.* **40**: 1295–9.

Augustin, J., McDole, R. E., McMaster, G. M., Painter, C. G. & Sparks, W. C. (1975). Ascorbic acid content in Russet Burbank potatoes. *J. Food Sci.* **40**: 415–16.

Augustin, J., Johnson, S. R., Teitzel, C., True, R. H., Hogan, J. M., Toma, R. B., Shaw, R. L. & Deutsch, R. M. (1978). Changes in nutrient composition of potatoes during home preparation. II. Vitamins. *Am. Potato J.* **55**: 653–62.

Akroyd, W. R. & Doughty, J. (1970). *Wheat in human nutrition*, FAO Nutritional Studies no. 23. FAO, Rome.

Brodribb, A. J. M. (1983). Dietary fibre as a tool of the clinician. In G. G. Birch & K. J. Parker (eds.). *Dietary fibre*. Applied Science Publishers, London and New York.

Burton, W. G. (1966). *The potato*, 2nd edn. Drukkerij Veenman BV, Wageningen.

Burton, W. G. (1974). Requirements of the users of ware potatoes. *Potato Res.* **17**, 374–409.

Cameron, M. & Hofvander, Y. (1976). *Manual on feeding infants and young children*, 2nd edn. Protein-Calorie Advisory Group, United Nations.

Caribbean Food and Nutrition Institute (1974). *Food composition table for use in the English-speaking Caribbean*. Caribbean Food and Nutrition Institute, Kingston, Jamaica.

Carter, E. G. A. & Carpenter, K. J. (1980). The availability of niacin in foods. *Federation Proceedings Abstrs.*, part I, Abstr. no. 1536, p. 557. Fed. Am. Socs. Exp. Biol., California.

Collazos, C. *et al.* (1974). [The composition of Peruvian foods.] In Spanish, 4th edn. Ministry of Health, Lima.

Doughty, J. (1982). Water, the hidden ingredient. *Appropr. Technol.* **8**: 11–12.

Dreher, M. L., Breedon, C. & Orr, P. H. (1983). Percent starch hydrolysis and dietary fiber content of chipped and baked potatoes. *Nutr. Rep. Int.* **28**: 687–91.

Englyst, H., Wiggins, H. S. & Cummings, J. H. (1982). Determination of the non-starch polysaccharides in plant foods by gas–liquid chromatography of constituent sugars as alditol acetates. *Analyst* **107**: 307–18.

Finglas, P. M. & Faulks, R. M. (1984a). Nutritional composition of UK retail potatoes, both raw and cooked. *J. Sci. Food Agric.* **35**: 1347–56.

Finglas, P. M. & Faulks, R. M. (1984b). The HPLC analysis of thiamin and riboflavin in potatoes. *Food Chem.* **15**: 37–44.

Finglas, P. M. & Faulks, R. M. (1985). A new look at potatoes. *Nutr. Food Sci.* no. 92: 12–14.

Fürer-Haimendorf, C. von (1964). *The sherpas of Nepal: Buddhist Highlanders*. University of California Press, Berkeley CA.

Griffiths, D. W. & Thomas, T. A. (1981). Phytate and total phosphorus content of field bean (*Vicia faba L.*). *J. Sci. Food Agric.* **32**: 187–92.

Gursky, M. J. (1969). A dietary survey of three Peruvian highland communities. M.A. thesis, Pennsylvania State University.

Hellendoorn, E. W., Noordhoff, M. G. & Slagman, J. (1975). Enzymatic determination of the indigestible residue (dietary fibre) content of food. *J. Sci. Food Agric.* **26**: 1461–8.

Jeltema, M. A. & Zabik, M. E. (1980). Revised method for quantitating dietary fibre components. *J. Sci. Food Agric.* **31**: 820–9.

Jones, G. P., Briggs, D. R., Wahlquist, M. L. & Flentje, L. M. (1985). Dietary fibre content of Australian foods. I. Potatoes. *Food Technol. Australia* **37**: 81–3.

Klein, L. B., Chandra, S. & Mondy, N. I. (1980). The effect of phosphorus fertilization on the chemical quality of Katahdin potatoes. *Am. Potato J.* **57**: 259–66.

Kubisk, A., Tomkowiak, J. & Andrzejewska, M. (1978). [The content of some trace elements in different parts of the tuber in five potato varieties.] In Polish. *Hodowla Rosl. Aklim. Nasienn.* **22**: 81–8.

Lampitt, L. H. & Goldenberg, N. (1940). The composition of the potato. *Chem. and Ind.* **59**: 748–61.

Leichsenring, J. M. et al. (1951). *Factors influencing the nutritive value of potatoes*, Minnesota Technical Bulletin no. 196. University of Minnesota Agricultural Experiment Station. Minnesota.

López de Romaña, G., Graham, G. G., Mellits, E. D. & MacLean, W. C. (1981). Prolonged consumption of potato based diets by infants and small children. *J. Nutr.* **111**: 1430–6.

McCay, C. M., McCay, J. B. & Smith, O. (1975). The nutritive value of potatoes. In W. F. Talburt & O. Smith (eds.), *Potato processing*, 3rd edn. AVI Publishing Co., Westport CT.

Mondy, N. I., Koch, R. L. & Chandra, S. (1979). Influence of nitrogen fertilization on potato discoloration in relation to chemical composition. 2. Phenols and ascorbic acid. *J. Agric. Food Chem.* **27**: 418–20.

Myron, D. R., Givand, S. H. & Nielsen, F. H. (1977). Vanadium content of selected foods as determined by flameless atomic absorption spectroscopy. *J. Agric. Food Chem.* **25**: 297–300.

National Advisory Committee on Nutrition Education (1983). *Proposals for nutritional guidelines for health education in Britain*. Health Education Council, London.

National Food Survey Committee (1983). *Household food consumption and expenditure, 1983*, Annual Report of the National Food Survey Committee. HMSO, London.

National Research Council (1980). *Recommended dietary allowances*, 9th edn. National Academy of Sciences, Washington DC.

Nyman, M., Siljeström, M., Pedersen, B., Bach Knudsen, K. E., Asp, N.-G., Johansson, C.-G. & Eggum, B. O. (1984). Dietary fiber content and composition in six cereals at different extraction rates. *Cereal Chem.* **61**: 14–19.

Orr, P. H., Toma, R. B., Munson, S. T. & D'Appolonia, B. (1982). Sensory evaluation of breads containing various levels of potato peel. *Am. Potato J.* **59**: 605–11.

Page, E. & Hanning, F. M. (1963). Vitamin B_6 and niacin in potatoes. *J. Am. Diet. Assoc.* **42**: 42–5.

Passmore, R., Nicol, B. M., Narayana Rao, M., Beaton, G. H. & De Maeyer, E. M. (1974). *Handbook on human nutritional requirements, WHO Monograph Series* no. 61. FAO/WHO, Geneva.

Paul, A. A. & Southgate, D. A. T. (1978). *McCance and Widdowson's The composition of foods*, 4th edn. MRC special report no. 297. HMSO, London.

Perloff, B. P. & Butrum, R. R. (1977). Folacin in selected foods. *J. Am. Diet. Assoc.* **70**: 161–72.

Pimentel, D., Dritschilo, W., Krummel, J. & Kutzman, J. (1975). Energy and land constraints in food production. *Science* **190**: 754–61.

Poats, S. V. & Woolfe, J. A. (1982). Feeding people with potatoes: the importance of nutritional considerations in potato research and acceptability. Seminar presented at International Potato Center. Mimeograph, Lima.

Quick, W. A. & Li, P. H. (1976). Phosphorus balance in potato tubers. *Potato Res.* **19**: 305–12.

Reistad, R. (1983). Content and composition of non-starch polysaccharides in some Norwegian plant foods. *Food Chem.* **12**: 45–59.

Salaman, R. N. (1949). *The history and social influence of the potato.* Cambridge University Press, Cambridge. (Reprinted 1970.)

Sandberg, A.-S. (1982). *Dietary fibre – determination and physiological effects.* Dept. of Clinical Nutrition and Dept. of Surgery II, University of Göteborg.

Schweizer, T. F. & Wursch, P. (1981). Analysis of dietary fiber, in W. P. T. James & O. Theander (eds.). *Basic and clinical nutrition*, vol. 3 *The analysis of dietary fiber in food.* Marcel Dekker, Inc., New York and Basel.

Spring, J. A., Robertson, J. & Buss, D. H. (1979). Trace nutrients. 3. Magnesium, copper, zinc, vitamin B_6, vitamin B_{12} and folic acid in the British household food supply. *Br. J. Nutr.* **41**: 487–93.

Swaminathan, K. & Pushkarnath (1962). Nutritive value of Indian potato varieties. *Indian Potato J.* **4**: 76–83.

Szkilladziowa, W., Secomska, B., Nadolna, I., Trzebska-Jeska, I., Wartanowicz, M. & Rakowska, M. (1977). Results of studies on nutrient content in selected varieties of edible potatoes. *Acta Aliment. Polon.* **3**: 87–97.

Teich, A. H. & Menzies, J. A. (1964). The effect of nitrogen, phosphorus and potassium on the specific gravity, ascorbic acid content and chipping quality of potato tubers. *Am. Potato J.* **41**: 169–73.

Theander, O. (1983). Advances in the chemical characterisation and analytical determination of dietary fibre components. In G. G. Birch & K. J. Parker (eds.). *Dietary fibre.* Applied Science Publishers, London and New York.

Thorn, J., Robertson, J. & Buss, D. H. (1978). Trace nutrients. Selenium in British food. *Br. J. Nutr.* **39**: 391–6.

Toma, R. B., Augustin, J., Shaw, R. L., True, R. H. & Hogan, J. M. (1978a). Proximate composition of freshly harvested and stored potatoes (*Solanum tuberosum* L.). *J. Food Sci.* **43**: 1702–4.

Toma, R. B., Orr, P. H., D'Appolonia, B., Dintzis, F. R. & Takekhia, M. M. (1979). Physical and chemical properties of potato peel as a source of dietary fibre in bread. *J. Food Sci.* **44**: 1403–1407.

Trowell, H., Southgate, D. A. T., Wolever, T. M. S., Leeds, A. R., Gassull, M. A. & Jenkins, D. J. A. (1976). Dietary fibre redefined. *Lancet* **1**, 967.

True, R. H., Hogan, J. M., Augustin, J., Johnson, S. R., Teitzel, C., Toma, R. B. & Shaw, R. L. (1978). Mineral composition of freshly harvested potatoes. *Am. Potato J.* **55**: 511–19.

True, R. H., Hogan, J. M., Augustin, J., Johnson, S. R., Teitzel, C., Toma, R. B. & Orr, P. (1979). Changes in the nutrient composition of potatoes during home preparation. III. Minerals. *Am. Potato J.* **56**: 339–50.

Walters, C. B., Sherlock, J. C., Evans, W. H. & Read, J. I. (1983). Dietary intake of fluoride in the United Kingdom and fluoride content of some foodstuffs. *J. Sci. Food Agric.* **34**: 523–28.

Watt, B. K. & Merrill, A. L. (1975). *Composition of foods: raw, processed, prepared*, Agriculture Handbook no. 8. U.S. Dept. of Agriculture, Washington DC.

Wenlock, R. W., Buss, D. H. & Dixon, E. J. (1979). Trace nutrients. 2. Manganese in British food. *Br. J. Nutr.* **41**: 253–61.

References

Werge, R. (1979). The potato and its potential role in the tropics. Mimeograph. International Potato Center, Lima.

WHO (1985). *Energy and protein requirements. Report of a joint FAO/WHO/UNU Expert Consultation*, WHO Tech. Rep. Ser. 724. WHO, Geneva.

Wills, R. B. H., Lim, J. S. K. & Greenfield, H. (1984). Variation in nutrient composition of Australian retail potatoes over a 12-month period. *J. Sci. Food Agric.* **35**: 1012–17.

Wolnik, K. A., Fricke, F. L., Capar, S. G., Brande, G. L., Meyer, M. W., Duane Satzger, R. & Kuennen, R. W. (1983). Elements in major raw agricultural crops in the United States. 2. Other elements in lettuce, peanuts, potatoes, soybeans, sweet corn and wheat. *J. Agric. Food Chem.* **31**: 1244–9.

Woodham-Smith, C. (1962). *The great hunger: Ireland 1845–1849*. Harper and Row, New York.

Wu Leung, W.-T. & Flores, M. (1961). *Food composition table for use in Latin America*. Institute of Nutrition of Central America and Panama, Guatemala/Interdepartmental Committee on Nutrition for National Defense, Bethesda, MD.

Wu Leung, W.-T., Busson, F. & Jardin, C. (1968). *Food composition table for use in Africa*. US Department of Health, Education and Welfare Public Health Service. Maryland, and FAO Nutrition Division, Rome.

Wu Leung, W.-T., Butrum, R. R. & Chang, F. H. (1978). *Food composition table for use in East Asia*, parts I & II. US Department of Health, Education and Welfare Publication no. (NIH) 79-465.

Zawadzka, T. (1978). [Contents of chromium in potatoes and in vegetable and fruit products.] In Polish. *Rocz. Panstw. Zakl. Hig.* **29**: 635–40.

3

Protein and other nitrogenous constituents of the tuber

As established in the previous chapter, potato is not a rich source of energy (approx. 335 kJ (80 kcal)/(100 g), but it supplies high-quality protein. This is of considerable importance in developing countries where energy supplies tend to be more readily available than protein. The nitrogenous constituents of the potato tuber have a high nutritional value compared with many other vegetable crops and there is a wealth of literature devoted to the subject.

Part 1 of this chapter addresses the factors affecting the composition and quality of tuber N and hence its contribution to the diet; Part 2 assesses ways of measuring the nutritional value. The last part discusses the possibilities for reclamation of valuable protein from waste processing. This may be of use to developing countries in planning potato processing operations.

Part 1: Composition of tuber nitrogen

Factors affecting total tuber nitrogen

The average contents of total protein in potato are approximately 2% (FWB) and 10% (DWB). Total protein is Kjeldahl N × 6.25, according to van Gelder (1981), although conversion factors of 5.7 and 7.5 have been suggested (Vigue & Li, 1975; Desborough & Weiser, 1974). Wide ranges of crude protein contents have been reported, e.g. 11.6% to 16.1% (DWB) between different species of *Solanum* and 9.5% to 14% (DWB) between different varieties of *S. tuberosum* (Hoff *et al.*, 1978; see also Espinola, 1979; Snyder & Desborough, 1980; International Potato Center, unpublished data). As the potato absorbs little water on boiling or steaming, the total protein content of boiled, unpeeled potato is virtually identical with that of the raw, uncooked tuber. Furthermore, Table 3.1 shows clearly that comparisons of raw potato with other uncooked vegetable sources, especially rice, are misleading; we should consider the protein content of the cooked foods.

Variations in total N of potato tubers are attributable not only to varietal differences, but also to cultivation practices, climatic effects, growing season and location. The increase of total N with increasing levels of applied N is well documented (Eppendorfer *et al.*, 1979; Hoff *et al.*, 1978; Mulder & Bakema, 1956; Rexen, 1976). Total N was increased significantly also by phosphorus applications of 56 kg/ha (Klein *et al.*, 1980). A study of the effects of moderate (22 °C/18 °C, day/night) and cool (11 °C/7 °C, day/night) temperatures on tuber protein metabolism showed that, in general, lower temperatures stimulated an increase in the percentage of tuber N (Vigue, 1973). Marked differences in total N have been found between varieties (Hunnius *et al.*, 1976; Peare & Thompson, 1975) and between crops from different years and locations (Li & Sayre, 1975; Hunnius *et al.*, 1976; Peare & Thompson, 1975). Breeding-programme selections for increased nitrogen content should therefore take place in conditions representative of the intended area of production.

Distribution of N within the tuber is not homogeneous (Neuberger & Sanger, 1942; Herrera, 1979), being highest in the skin, decreasing in the cortex and rising again towards the pith. Desborough & Weiser (1974) reported that protein nitrogen content was similar in the cortical, medullary and pith regions, whereas the pith area was found to have significantly more non-protein nitrogen (NPN) than did the cortex tissue (Ponnampalam & Mondy, 1983).

Table 3.1. *Crude protein content of raw and cooked potato and rice*[a]

Food	Moisture (%)	Total protein (N × 6.25) (%)
Raw rice	12.0	6.7
'Dried' potato[b]	11.7	8.4
Raw potato	79.8	2.1
Cooked potato[c]	79.8	2.1
Cooked rice[d]	72.6	2.0

[a] From Watt & Merrill (1975).
[b] See note (*d*) Table 2.3.
[c] Boiled in skin.
[d] Boiled, milled white.

Constituents

The total N of potato tubers comprises: (1) soluble, coagulable (true) protein; (2) insoluble protein; and (3) soluble NPN, which is composed of free amino acids, the amides asparagine and glutamine, and small amounts of nitrate N and basic nitrogen compounds including nucleic acids and alkaloids. The insoluble protein fraction occurs mainly in the peel; in a rat feeding experiment, the removal of the skin and outer cortex increased the nutritive value of the remainder of the potato (Chick & Slack, 1949).

Table 3.2 shows the average content of the various soluble fractions of potato N. The distribution of the main nitrogenous fractions of total tuber N in each of four varieties is given by Miedema *et al.* (1976), who showed that insoluble N is only about 4% of total N.

The proportion of soluble protein N to total N, although given as 50% in Table 3.2, can vary widely. A range of 29.5% to 51.2% was noted among 11 different *S. tuberosum* varieties (Neuberger & Sanger, 1942) and of 40% to 74% in 50 clone samples of *S. tuberosum* group *andigena* (Li & Sayre, 1975).

Table 3.2. *Fractions of the soluble nitrogen of potato*[a]

Nitrogen fractions	% of total N
True protein N	50
NPN	50
Free amino acid N	15
Asparagine N	13
Amide N	
Glutamine N	10
Basic N	8[b]
Nitrate N	1
Nitrite N	Trace
Ammonia N	3[c]

[a] Based on data from Knorr (1978).
[b] Alkaloids, certain vitamins, purines, pyrimidines, quaternary ammonium compounds, etc.
[c] It is doubtful whether there is so much free ammonia in potatoes; probably part of it is formed from glutamine, which is readily hydrolysed during analysis.

Soluble protein

True potato protein contains substantial levels of the essential amino acids. Lysine content is particularly favourable, being comparable with that of whole egg (van Gelder & Vonk, 1980).

An analysis of coagulable freeze-dried protein from 34 varieties of potato showed that, although the varieties covered a range of coagulable protein from 0.37 to 1.24 g/100 g fresh tuber, there was little variation in amino acid composition among them (van Gelder & Vonk, 1980). There was no correlation between coagulable protein content and methionine, lysine or the essential amino acid (EAA) index, which ranged from 86 to 93. The authors contrasted potato with (1) barley, where protein content shows a negative relation with protein lysine and (2) wheat, where protein content is negatively related to content of lysine and of methionine. They concluded, therefore, that breeding for higher levels of true protein will not adversely affect the nutritional value of that protein. There was little variability in the amino acid composition of the true protein of seven varieties of *S. tuberosum*, and high values for methionine were not found amongst 45 wild species of *Solanum* (Hoff *et al.*, 1978). Wide ranges in the contents of true protein essential amino acids were encountered when 40 genotypes were analysed (Desborough & Weiser, 1974); for example, 0.19 to 2.69 and 1.62 to 10.82 mg/g dry tuber for methionine and lysine, respectively. Highly significant positive correlations were found between contents of protein-bound essential amino acids and protein, being particularly high for leucine, lysine and phenylalanine. The authors concluded that selection for amino acids such as methionine that are present in relatively low amounts could be beneficial.

The amino acid composition of the true protein of a particular variety of potato is genetically determined and is little affected by environmental conditions (Eppendorfer *et al.*, 1979). Mulder & Bakema (1956) found that the amino acid composition of potato protein was independent of the mineral nutrition of the plants in the case of nitrogen, phosphorus and potassium supplies. Tjørnholm *et al.* (1975) demonstrated that true protein content was increased by nitrogen fertilization, but that the amino acid composition of the protein was not influenced by nitrogen and potassium supplies. However, mineral nutrition may have an effect on the content and proportion of protein present, even though the composition remains unchanged. The percentage of true protein in total protein is subject to environmental effects. Some authors (Mulder & Bakema, 1956; Labib, 1962) have found that the ratio protein N : total N is lower with high nitrogen fertilization than with low. In contrast, others (Hoff *et* al., 1978; Li & Sayre, 1975) have shown that increasing nitrogen fertilization increases not only the percentage of total N, but also the percentage of true protein, and that true protein and total N increase at a similar rate (Hoff et al., 1978). It has been suggested that, in screening tubers for protein content, it is the soluble, coagulable protein which should be determined and efforts should be made to breed for higher levels of this protein (Hoff et al., 1978).

Non-protein nitrogen

Kapoor et al. (1975) found that most of the NPN fraction (75%) is in the form of free amino acids and amides. Free amino acids constitute 22% to 35% of the total tuber amino acid content, whilst the amides, asparagine and glutamine, are present in about equal amounts and together comprise approximately half the total free amino acids (Hoff et al., 1978). Synge (1977) has reviewed the contents of different free amino acids in 13 varieties of potatoes grown in seven countries and has shown wide ranges.

Most of the amino acids that occur in true protein have also been found in the NPN. Kapoor et al. (1975) did not find tryptophan in the free form, but other authors have found it in NPN (Synge, 1977). Cystine/cysteine has been reported as absent from the NPN (Hoff et al., 1971; Kaldy, 1971; Tjørnholm et al., 1975), absent completely in some varieties and present as traces in others (Herrera, 1979). In general, the NPN contains lower amounts of the essential amino acids than does true protein (Labib, 1962; Kapoor et al., 1975; see Table 3.3). However, Kaldy (1971), in an analysis of six varieties, found that, on average, methionine was higher in the NPN fraction in comparison with the value in the whole tuber and concluded that free methionine made a significant contribution to the protein quality of potato. Free methionine accounted for 93% of the variations in available methionine in the potato clones analysed by Luescher (1972). It ranged from 0.34 to 2.07 mg/16 mg NPN and contributed 12% to 62% of all the methionine present in the total N. There appears to be no relation between the relative proportions of free and protein-bound amino acids (Hoff et al., 1978; Thompson & Steward, 1952).

In contrast to soluble protein, the amino acid composition of the NPN fraction is extremely variable and subject to the influence of factors such as mineral nutrition, variety, soil, and climatic conditions.

The positive effects of nitrogen application upon the content of soluble NPN have been emphasized (Mulder & Bakema, 1956; Tjørnholm et al., 1975). Moderate application of phosphorus also significantly increased NPN (Klein et al., 1980), as did a deficiency of either potassium or magnesium (Mulder & Bakema, 1956). In that application of N and phosphorus, on the one hand, and potassium on the other, have opposite effects upon the nitrogen content in the tubers, the overall effect of application of all three major nutrients is unpredictable (Burton, 1966). However, the content of free amino acids has been raised by increased applications of a combination of N, phosphorus, potassium and calcium (Tikhonov & Bychkov, 1969).

An increase in the NPN fraction does not lead to an equal increase in all free amino acids and some may decrease. In general, the literature shows that the greatest increases occur in the amides. Mulder & Bakema (1956) assumed that glutamine, asparagine and arginine are the forms in which absorbed N accumulates in the tuber before being further metabolized. One would therefore expect an alteration in conditions affecting nitrogen accumulation to have the greatest affects on levels of these compounds.

There are reports of variable responses of free amino acids, apart from the amides, to external influences. Mulder & Bakema (1956) found that a high level of NPN in total N correlated with a high NPN amide level, but that other free amino acids generally decreased. Labib (1962) observed that immediately after harvest the content of all essential amino acids in the NPN fraction was inversely proportional to the nitrogen fertilization. Eppendorfer et al. (1979) found that, in NPN, the percentage of free

Table 3.3. *Relative proportions of essential amino acids in true protein and NPN found in one variety*[a]

Amino acid	Protein[b]	NPN[c]
	(mg/g dried potato powder)	
Histidine	1.42	0.60
Isoleucine	2.80	1.28
Leucine	5.90	0.74
Lysine	4.40	0.53
Methionine	1.23	0.26
Phenylalanine	2.46	0.85
Tyrosine	2.52	2.25
Threonine	3.19	0.89
Tryptophan	0.86	0.00
Valine	3.56	2.60

[a] Calculated from figures of Kapoor et al. (1975).
[b] N not extracted by shaking dried potato powder with ethanol for 30 min is 53% of total N.
[c] N extracted with ethanol is 47% of total N.

amino acids, other than aspartic and glutamic acids and their amides, decreased with increasing percentage of total N in the dry matter. On the other hand, Tjørnholm et al. (1975) increased the content of each of the free amino acids by increased application of N, aspartic and glutamic acids and their corresponding amides responding most. According to Hoff et al. (1971), the response of individual amino acids to increased nitrogen fertilization varied considerably, the greatest increase occurring in glutamic acid (and glutamine), but with tyrosine decreasing fractionally.

In several varieties of potatoes grown in England and Ireland, soil and climatic conditions influenced the concentrations of some free amino acids more than others (Davies, 1977). Levels of glutamine, alanine, valine and tyrosine were higher and proline and histidine lower in samples from Ireland, where rainfall and humidity are greater, than in those from England. When low temperatures stimulated an increase in tuber N, the greatest increases in N were noted in glutamic acid, proline and arginine (Vigue, 1973).

It has been suggested that, nutritionally, the relatively large amounts of glutamine present may be significant (Chick & Slack, 1941). Glutamine and glutamic acid are important in transamination reactions (Smith, 1968). Chick (1950) reported that the amino acids of potato NPN supplement those of wheat gluten, thereby being able to increase the nutritional value of the latter. Potato free amino acids are 100% available for absorption and the percentage of NPN present in total N may influence the overall digestibility of that nitrogen.

Amino acid composition of the whole tuber

From the variations in total N and true protein and NPN fractions discussed above, it is clear that there can be considerable differences in the amino acid composition of the whole potato as eaten. It has been shown that increasing levels of total N in dry matter (DM), brought about by increased nitrogen fertilization, generally result in decreasing protein quality in terms of amino acid concentrations as g/16 g N (i.e. g/100 g protein: Rexen, 1976; Eppendorfer et al., 1979). The reasons for this are not given, but may be linked partly to a decrease in true protein content as a percentage of total N and also to an increase in amide N at the expense of the essential amino acids in the free form.

However, Rexen (1976) noted a great difference in the behaviour of different varieties in this respect. Some varieties showed a tendency for improved protein quality with increasing nitrogen content. It is possible that the percentage of true protein in these varieties actually increased with increasing total N. Moreover, his data show that amino acid contents in the whole tuber as eaten do not decrease with increasing nitrogen content, as they depend not only on the amino acid composition of the N but also on the quantity of total N in the DM and hence the percentage of DM in the tuber (see Table 3.4). Talley (1983) has also shown this by reinterpreting the data of Eppendorfer et al. (1979) and recent work by Millard (1986) demonstrated an improvement in the protein quality of one variety on a fresh weight basis with increasing % N in DM resulting from increasing levels of fertilizer application. In each of three experiments (using the same variety, grown at three different locations in Scotland) there were statistically significant increases in the percentages of the adult daily requirements supplied by each of the essential amino acids in 100 g fresh tuber, as nitrogen application was increased from 0 to 250 kg N/ha. The only exceptions were the sulphur-containing amino

Table 3.4. *Amino acid composition of the total N and of the tuber, at two different levels of total N*[a]

	Level 1			Level 2		
	g/16 g N	% in DM	% in fresh tuber	g/16 g N	% in DM	% in fresh tuber
Total N	—	1.23[b]	—	—	1.50[c]	—
DM	—	—	28.10[b]	—	—	26.99[c]
Histidine	1.93	0.15	0.04	1.79	0.17	0.05
Isoleucine	3.49	0.27	0.08	3.34	0.31	0.08
Leucine	5.59	0.43	0.12	5.51	0.52	0.14
Lysine	5.94	0.42	0.12	5.27	0.49	0.13
Methionine	1.85	0.14	0.04	1.68	0.16	0.04
Phenylalanine	3.94	0.30	0.08	3.77	0.35	0.09
Tyrosine	3.35	0.26	0.07	3.22	0.30	0.08
Threonine	4.06	0.31	0.09	4.25	0.40	0.11
Valine	4.70	0.36	0.10	4.54	0.43	0.12
Alanine	3.68	0.28	0.08	3.12	0.29	0.08
Arginine	4.20	0.32	0.09	4.19	0.39	0.11
Aspartic acid	20.46	1.57	0.44	21.71	2.03	0.55
Glutamic acid	13.36	1.03	0.29	13.42	1.26	0.34
Glycine	3.28	0.25	0.07	3.18	0.30	0.08
Proline	3.82	0.29	0.08	3.77	0.35	0.09
Serine	4.25	0.33	0.09	4.16	0.39	0.11

[a] Data from Rexen (1976). Means of 33 varieties.
[b] Tubers resulting from fertilization with 114 kg N/ha.
[c] Tubers resulting from fertilization with 186 kg N/ha.

acids, in one of the three experiments, which increased but not significantly. It is obviously valuable to compare the protein quality of potatoes in general with the proteins of other foodstuffs, which have greatly differing nitrogen contents, in terms of g/16 g N. However, it is also useful to compare the protein nutritive value of different tuber samples in terms of the food as eaten. This will depend on the combination of a number of factors: the amino acid composition of the total N (influenced by the ratio protein N:NPN), the total N content of the tubers as eaten and the availability (digestibility) of the N (which is partly dependent on the concentration of free amino acids). These are influenced not only by genetic factors but also by environmental conditions, cultivation practices and interactions between these variables.

Part 2: Nutritive value of tuber nitrogen

The evaluation of potato protein (true protein and free amino acids) has been carried out more frequently by chemical amino acid analyses, microbiological growth assays and animal feeding tests than by feeding experiments with humans. The relatively rapid chemical and microbiological analyses, normally used for initial monitoring, have generally indicated that potato protein is of good quality. Animal feeding tests have not shown consistent results: this is not surprising as workers have used different concentrations of potato protein in diets fed to rats (Knorr, 1978), and PER determinations are notoriously difficult to reproduce between laboratories. Feeding trials with humans, which are complex and costly to carry out, and therefore less frequently employed, have confirmed the high quality of potato protein for human consumption.

Amino acid analyses and scores

There are few complete amino acid analyses of potato tubers in the literature. Kaldy & Markakis (1972) and Knorr (1978) provide tables of previous analyses by several authors, and more recently Talley *et al.* (1984) reported the amino acid compositions of several North American varieties grown in different locations. Eppendorfer *et al.* (1979) determined the amino acid compositions of potato samples with varying levels of N in their DM. Table 3.5 shows the essential amino acid composition of raw potato given in four different sources. These confirm the high levels of lysine in potato protein.

Concentrations of potato amino acids have been variously reported in the literature, on the basis of g/16 g N, mg/g N, mg/g DM, μmol/g DM and mg/g food. There is a need for standardization of reporting, to avoid confusion and facilitate comparisons of the nutritive value of different samples. It should also be remembered that meaningful comparisons of the capacity of different potato samples to satisfy protein needs must be made on the basis of the amino acid concentrations in the food as eaten. These will depend on the composition and on the content of tuber N.

The usual method of reporting tuber amino acid composition (g/16 g N) has been used to evaluate the nutritive value of potato as compared with the amino acid pattern of whole egg (Table 3.5). The protein scores of six potato varieties analysed by Kaldy & Markakis (1972) varied from 60 to 78, on the basis of the sulphur-containing amino acids. Methionine was found to be the first limiting amino acid. Rexen (1976), however, found that in some cultivars the limiting amino acid was isoleucine. He reported that, in relation to whole egg EAA indices of 33 potato varieties ranged from 55 to 84. For some samples of high protein potato hybrids, Desborough & Lauer (1977) found a range of EAA indices from 72 upwards; one sample scored 100, and was therefore equal to egg protein.

Table 3.5. *Essential amino acid composition of potato as reported by various authors (g/16 g N)*

Amino acid	A	B	C	D	Average	Whole egg (E)
Histidine	2.3	1.9	1.9	—	2.0	2.4
Isoleucine	4.2	3.4	4.2	3.6	3.9	5.6
Leucine	5.6	5.6	6.1	6.3	5.9	8.3
Lysine	6.2	5.6	5.4	6.7	6.0	6.2
Methionine	1.6 ⎫ 2.9	1.8	1.6 ⎫ 2.9	1.2 ⎫ 3.1	1.5 ⎫ 3.0	5.0
Cystine	1.3 ⎭	—	1.3 ⎭	1.9 ⎭	1.5 ⎭	
Phenylalanine	4.5 ⎫ 8.2	3.9 ⎫ 7.2	4.3 ⎫ 7.3	4.5 ⎫ 8.3	4.3 ⎫ 7.8	9.1
Tyrosine	3.7 ⎭	3.3 ⎭	3.0 ⎭	3.8 ⎭	3.5 ⎭	
Threonine	3.8	4.2	3.8	3.8	3.9	4.0
Tryptophan	1.5	—	1.4	1.2	1.4	1.0
Valine	5.7	4.6	5.1	5.0	5.1	5.0

A, Kaldy & Markakis (1972). Average of six varieties grown in North America.
B, Rexen (1976). Average of 33 varieties grown in Denmark.
C, Paul & Southgate (1978). Food composition tables, based on varieties grown in Britain.
D, López de Romaña et al. (1981b). One variety grown in Peru.
E, WHO (1973).

It has been indicated recently (WHO, 1985) that the amino acid scoring procedure, by which the capacity of a protein or mixture of proteins to meet the essential amino acid and N requirements of a human recipient is evaluated, must be based on a knowledge of those requirements. The latest estimates of patterns of amino acid requirements (mg/g protein) for various age groups (WHO, 1985) are compared, in Table 3.6, with the concentrations in egg, cow's milk, beef and potato. For infants, potato protein does not satisfy most of the amino acid requirements, which are based on human milk, but it has a very high average amino acid score of 90 for the pre-school child and scores of over 100 for all other age groups. Potato protein has a particularly favourable lysine content in comparison with cereal proteins (see Table 2.5, p. 30), whose amino acid scores, on the basis of human requirements, are much lower.

In fully evaluating the capacity of potato and potato-based diets to provide safe levels of protein intake in humans, especially children, digestibility must be taken into account. WHO (1985) recommends the calculation of safe levels of dietary protein intake in different age groups as follows:

Safe level of dietary protein (e.g. in potato or potato-based diet) for age group = Safe level of reference protein for age group × 100/(Amino acid score of dietary protein for age group) × 100/(Digestibility of dietary protein)

The quantity of potato or of a potato-based diet needed to satisfy this calculated safe protein intake then depends on its protein content as eaten.

More information is required about the degree of potato protein digestibility, either alone or in mixed diets. The little knowledge available suggests that it may be rather low. The high amino acid score, however, is reflected by the high nutritive value of potato protein which has been found in practice (see the section on human feeding experiments, below).

Microbiological assays, animal feeding and digestibility experiments

In 1962, Labib used a *microbiological assay* with *Tetrahymena pyriformis* W. to assess the protein nutritive value of four varieties of potato. They averaged 89 when related to casein (100) and 75 when related to whole egg protein (100). Luescher (1972) assessed the biological value (BV) of total N and NPN in 16 clone samples to be 84 and 78,

Table 3.6. *Comparison of suggested amino acid requirement patterns with amino acid composition of potato*

Amino acid (mg/g crude protein)	Suggested pattern of requirement					Reported composition			
	Infant		Preschool child (2–5 yr)	School child (10–12 yr)	Adult	Egg	Cow's milk	Beef	Potato
	Mean	(Range)							
His	26	(18–36)	19	19	16	22	27	34	20
Ile	46	(41–53)	28	28	13	54	47	48	39
Leu	93	(83–107)	66	44	19	86	95	81	59
Lys	66	(53–76)	58	44	16	70	78	89	60
Met + Cys	42	(29–60)	25	22	17	57	33	40	30
Phe + Tyr	72	(68–118)	63	22	19	93	102	80	78
Thre	43	(40–45)	34	28	9	47	44	46	39
Tryp	17	(16–17)	11	9	5	17	14	12	14
Val	55	(44–77)	35	25	13	66	64	50	51
Total									
Incl. his	460	(408–588)	339	241	127	512	504	479	382
Excl. his	434	(390–552)	320	222	111	490	477	445	363

Data from World Health Organization (1985).

Amino acid score = $\dfrac{\text{mg amino acid in test protein}}{\text{mg amino acid in requirement pattern}} \times 100$.

Amino acid scores of potato: Infant = 63; Preschool child = 90; School child/Adult = > 100.

respectively, using *Streptococcus zymogenes* as the test organism. Meister (1977) also using *S. zymogenes* found that the BV of potato protein ranged from 77 to 82.

Animal feeding experiments with potato are more numerous: (for a review, see Markakis, 1975). In some cases no attempt was made to fulfil the dietary requirements for all nutrients other than protein, so that the quality of the diet was limited by factors additional to protein quality. Such experiments have produced poor growth in rats when potato was fed as the major source of protein in the diet (Joseph *et al.*, 1960; Roy Choudhuri *et al.*, 1963). However, they have demonstrated that potato can supplement cereal proteins. Rats grew significantly better than rice-only fed controls when 25% of the rice in an Indian rice-based diet was replaced by potato (Joseph *et al.*, 1960). However, Roy Choudhuri *et al.* (1963) found that this supplementation effect could only be demonstrated when the diets contained an adequate calcium supply. Early experiments by McCollum *et al.* (1918, 1921) showed that mixtures of several cereals (including wheat, oats and maize) with potatoes produced better rat growth than did the cereals alone.

Chang & Avery (1969) showed that the nutritive value of potato protein was superior to that of rice for weanling rats although they were actually comparing cooked potato with raw rice. Chick & Slack (1949), observing that potato protein promoted good growth in young albino rats, reported a protein efficiency ratio (PER) of 1.8 for potato proteins at a 10% level of protein intake. Joseph *et al.* (1963) found that the variety which produced the highest PER (1.99) in growing rats also contained greater amounts of the sulphur-containing amino acids than did two other varieties tested. The average PER for a number of high protein potato hybrids given in the form of potato flakes was 2.3 (Desborough *et al.*, 1981), nine of the genotypes tested having protein equal in quality to that of casein.

Little is known about the digestibility of potato protein. Values for apparent digestibility (i.e. digestibility not corrected for metabolic faecal N losses) in rats have varied from about 60% to about 80% (Chick & Slack, 1949; Espinola, 1979). The true (i.e. corrected) digestibility of protein in freeze-dried boiled potato, determined with rats, varied from 82.7% for a sample containing 1.4% total N in DM to 90.8% in one with 3.07% N in DM (Eppendorfer *et al.*, 1979). The authors suggest that the increase in digestibility might be due to a lower proportion of the N being incorporated into fibrous substances in potatoes with higher total N. An alternative explanation, however, might be that a greater proportion of the N in high-nitrogen potatoes was in the form of fully available free amino acids (López de Romaña *et al.*, 1980).

A protein digestibility test *in vitro* using mammalian proteolytic enzymes was applied to potato flakes prepared from a number of potato hybrids (Boody & Desborough, 1984). The average digestibility *in vitro* was 72.4% (range 71.1% to 74.9%). There appears to be considerable variation in potato protein digestibility as evidenced by the above results, and the reasons for this and their applicability to human feeding need further investigation.

Human feeding experiments

Protein is needed by humans for growth (particularly in infants), maintenance of tissues, restoration of losses caused by damage or disease, and pregnancy and lactation. Different proportions of amino acids may be required for maintenance and for growth (Bender, 1982), and a particular protein may be more effective for one purpose than for another. Hence, feeding experiments with adults and children will be dealt with separately.

In normal meals, it should be noted, potatoes are rarely eaten alone as the sole source of N and there are supplementary effects as a result of mixing potatoes with other foods.

Adults

There are brief reviews of adult human feeding experiments with potatoes by Knorr (1978) and by Markakis (1975). In contrast to animal feeding trials, work with human adults has consistently shown that men or women can be maintained in nitrogen equilibrium and good health on diets in which all the N, or almost all, was supplied by potatoes.

Rose & Cooper (1917) maintained a young woman in nitrogen balance for seven days on an intake of 0.096 g nitrogen/kg body weight from potato. Kon & Klein (1928) kept a man and woman in nitrogen equilibrium and good health for almost six months on a diet in which all the N required was supplied by potatoes, the daily need for potato protein being 36 g for the man and 24 g for the woman on a 70 kg body weight basis.

Kofrányi & Jekat are cited by various authors (Herrera, 1979; Knorr, 1978; Markakis, 1975) as having determined that the average amount of protein necessary for the maintenance of nitrogen balance in the case of three healthy college students was 0.545 g/kg body weight for potato and 0.505 g/kg body weight for egg protein. Furthermore, in terms of quantities required to maintain nitrogen balance in adult human beings, the protein of potatoes had better nutritive value than the protein of beef, tuna fish, wheat flour, soybean, rice, corn, beans or seaweed. Thompson (1977) reported that the minimum protein requirement of a graduate student on a diet in which 95% of the protein was supplied by potatoes, was 0.518 g potato protein/kg body weight, which agrees with the values reported for two subjects by Kofrányi *et al.* (1970). It should be noted that the potato sample used for Thompson's experiment had a sulphur amino acid content 50% higher than average reported values.

Minimum protein requirement for nitrogen balance in two subjects, for a mixture of 36% egg N + 64% potato N, was lower than for egg or potato alone (Kofrányi *et al.*, 1970). This, Kofrányi (1973) suggested, was because the BV of a protein source is related to its overall amino acid pattern, rather than to the absolute amounts of essential amino acids present.

Animal experiments do not support these findings. Herrera (1979) measured the PER in growing rats of a 65:35 (protein) ratio of a mixture of dried potato and dried egg, and states that the PER of the potato–egg mixture was 8% higher than that of dried whole egg. However, his data show no statistically significant difference between the PERs of the mixture and whole dried egg. In another study with rats Eggum *et al.* (1981) could not substantiate Kofrányi's findings with humans, noting a curvilinear decrease in BV with increasing concentration of potato protein in an egg protein diet. These may simply reflect differences between species and their stage of maturity.

Kies & Fox (1972) reported that the addition of L-methionine to potato flakes (which provided 80% of the dietary N) in the diets of young adults improved the potato protein quality. The dehydrated 'instant' mashed potato flakes involved might have suffered some destruction of methionine during processing (see Chapter 4) as the authors' values of 1.12 and 2.4 g/16 g N for methionine and total sulphur amino acids, respectively, were somewhat lower than those for fresh potatoes given in Table 3.5.

Children

The only published work evaluating the quality of potato protein in infant diets is that by López de Romaña and co-workers (1980, 1981*a,b*) in Peru. Studies with infants and young children recovering from malnutrition demonstrated that potatoes can be used to supply all the daily dietary requirement for protein and a substantial part of that for energy.

A short-term experiment (López de Romaña *et al.*, 1980) assessed the digestibility and utilization of protein in diets providing only potato protein and at marginal levels of intake. The nitrogen balance of 11 children between the ages of 8 and 35 months, while consuming two diets in which potato provided approximately 5% of the energy as protein was compared with that of children consuming an isonitrogenous casein control diet. A first batch of dehydrated potatoes containing 5.75 g crude protein/100 g was incorporated into the diet so as to provide 75% of the total dietary energy. A second batch containing 9.13 g crude protein/100 g supplied all the dietary N but at a level that provided only 50% of the dietary energy. Apparent nitrogen retentions on the potato diets were significantly lower than those from the casein control diet. All differences in apparent nitrogen retention were due to inferior apparent nitrogen absorption, with no significant difference in apparent retention of absorbed nitrogen. This suggested that inferior digestibility of the potato protein rather than amino acid deficiency was responsible for its failure to match casein as a protein source for small children. The high amino acid scores of potato for small children shown in Table 3.6 would tend to confirm this.

The 'apparent BVs' (apparent nitrogen retention as a percentage of absorbed nitrogen) of potato protein and casein were 49% and 53%, respectively. Potato protein digestibility (expressed as a percentage of that of casein) was 79% and 92%, respectively, for the diets containing the lower- and higher-nitrogen potato batches.

The percentages of protein energy in the two test diets were equal, but the energy provided by the lower- and higher-nitrogen batches of potatoes was 75% and 50%, respectively. More potato starch must therefore have been present in the lower-nitrogen batch potato diets. Dreher *et al.* (1984) noted that most root/tuber starches have been found to be inferior to cereal starches with respect to their effect on protein utilization. Dreher *et al.* (1981) showed that PERs, net protein ratios and protein digestibilities in mice were lower with autoclaved potato starch as the major source of energy in the diet than with other more digestible autoclaved starches. López de Romaña *et al.* (1980) demonstrated that, as potato provided increasing percentages of dietary energy in small children, there were increasing stool weights and increasing losses of carbohydrate and energy in the faeces, reflecting poor digestibility of carbohydrate. The greater quantity of potato starch in the lower-nitrogen batch potato diet may therefore have had an adverse effect on potato protein digestibility. This could help to explain the authors' finding that the potato protein from the lower-nitrogen batch was no better digested than rice protein (digestibility 78% that of casein) in spite of the high free amino acid content of potato protein. The increased percentage of free amino acids they reported in the higher-nitrogen potato diet, together with its lower potato starch content, may account for its superior digestibility. Such relationships should be further investigated.

Another study (López de Romaña et al., 1981a) showed that infants and small children can consume from 50% to 75% of their energy and up to 80% of their nitrogen requirements as potato during a longer period (three months). Limiting factors for the consumption in diets containing increasing quantities of potato are the substantial bulk to be eaten and the relatively poor digestibility of potato carbohydrate. Hence, protein intake from lower-protein potatoes could not be improved by increased consumption. The protein status of the children, as determined by serum albumin concentrations, was maintained during the whole study, reflecting the adequacy of the dietary protein. Since potato is never the sole source of N in children's diets, the remaining protein should be provided by a less bulky food that is a source of high quality protein, not to improve the quality of the potato protein but to decrease the bulk and volume of the diet.

Further research (López de Romaña et al., 1981b) confirmed that potato protein has an adequate ratio of total essential amino acids to total amino acids and a balance among individual essential amino acid concentrations to meet the needs of infants and small children, if the protein is given and absorbed sufficiently to fulfil total nitrogen requirements.

Potato protein can be a useful weaning food in developing countries, especially if varieties with higher protein contents and better carbohydrate digestibilities are used. The experimental production of weaning foods containing potato has been reported (Abrahamson, 1978; Kaur & Gupta, 1982).

Comments on protein contribution from potatoes

Potato protein is of sufficiently high quality for maintenance purposes in adult man and for growth of infants and children. The relatively low digestibility of potato protein is a disadvantage when potatoes are used for feeding to children; potatoes have to be consumed in large quantities to satisfy both protein and energy requirements, a characteristic they share with other root and tuber staples. Potatoes are rarely consumed as the sole source of N in the diets of either adults or children, but it is clear that they can make a valuable contribution to the protein content and quality of a mixed diet, provided present levels of protein in potato are maintained. In searching for higher-yielding varieties, maintenance of protein levels in potatoes should not be overlooked by plant breeders.

Part 3: Potato protein from processing waste

Reasons for waste production

Large-scale potato-processing industries of the developed world produce great quantities of liquid waste effluents during the production of starch, flakes, granules, french fries (chips) and chips (crisps). Such waste effluents contain organic materials, have a high biological oxygen demand and constitute a serious pollution problem if discharged directly into the environment; their treatment at sewage disposal plants is costly.

Potato fruit water, one of the wastes from potato starch factories, contains about 2% (w/v) soluble solids, including protein, non-polymeric nitrogen compounds such as amino acids, and sugars and minerals. Interest in the recovery of potato protein from waste effluents has increased during the past 60 years (De Noord, 1976). Knorr (1977) reported two benefits, suggested by Heisler et al. (1959), of recovering nitrogen compounds from potato starch factory waste water: reduced pollution from protein or fruit water, and improved economics of starch production by the creation of marketable by-products. World production of potato starch amounts to approximately two million tons (1 ton = 1.016 tonne) (Meuser & Smolnik, 1979). About 100 000 tons of potato protein could be recovered if all potato starch producers were to install protein recovery systems. The introduction of stringent pollution control regulations on waste effluents in some countries and improved starch extraction processes leading to more concentrated potato fruit water (about 4% soluble solids; Strolle et al., 1980) have recently made protein recovery increasingly attractive to manufacturers. The 1970s also saw the initiation of research into the recovery of protein from the waste water of factories producing french fries and chips. In these industries, waste water results from peeling, cutting and blanching. The high biological oxygen demand of this water enables it to be used as a culture medium for the production of 'single-cell protein', e.g. fungal protein (Dambois et al., 1978; Abouzied & Mostafa, 1984) appropriate for animal feeding, or incorporation into food for humans (Skogman, 1976). By direct recovery of protein, a typical North American chip plant processing 31 tonnes potatoes/day, could also produce per day an estimated 550 kg dried protein concentrate containing 30% of soluble potato protein (Meister & Thompson, 1976a).

Protein recovery

About 60% of the solids in potato fruit water from starch factories is present as nitrogenous substances, half of which can be coagulated (precipitated) and hence recovered. The most common way to coagulate the protein is by heating, with or without pH adjustment, or by a combination of heat and pressure. Coagulation is followed by separation using filters, gravity settling or centrifugation. The separated protein concentrate is then dried, preferably by spray-drying, which yields products with a better colour and solubility than do methods such as tray- or drum-drying. Freeze-drying yields a soft, white product, but is expensive and therefore probably uneconomical (Strolle *et al.*, 1973). Methods for the recovery of protein concentrates from potato fruit water have been reviewed by Knorr (1977). More recently, Wojnowska *et al.* (1981) have explored the possibilities of using such methods as ultrafiltration (which yielded promising results), cryoconcentration and polyelectrolyte coagulation for protein recovery before subsequent drying. Knorr (1980) has investigated the use of various coagulants for utilization at room temperature as possible alternatives to the more costly use of energy in heat coagulation. The further recovery of by-products from the residual water, after protein coagulation and removal, has also been reviewed by Knorr (1977). This residual water also could be used as a growth medium for single-cell protein (Hutterer, 1978).

Nutritional value

Meuser & Smolnik (1979) stated that directly precipitated protein from potato fruit water is obtained with a purity of 80% to 85%. The typical composition of a commercial European potato protein concentrate (PPC), used for animal feed, contained 81.5% crude, and 79.7% pure, protein (Knorr, 1977).

The nutritional value of recovered protein is high, although dependent both upon the protein quality of the potatoes from which it is derived and on the method of recovery (Meister & Thompson, 1976b). The BV of recovered protein was higher than that of the crude protein of the original tubers (Meister & Thompson, 1976*b*), and its content of essential amino acids greater than that in the potato fruit water (Wojnowska *et al.*, 1981). Knorr (1978) gave the amino acid composition of some commercial PPC products and of PPCs coagulated by various methods; PPC has a high lysine content, and favourable levels of methionine and cystine compared with published values for the sulphur-containing amino acids in potato tubers. The PERs of two commercial heat-coagulated PPCs were not significantly different from that of casein (Knorr, 1978).

Use in food for humans

Attempts to utilize PPCs, without further purification, for human consumption by enriching bakery products, such as crispbread, cookies, crackers, wafers and biscuits (Meuser & Smolnik, 1979), were limited in success due to the undesirable taste, smell and texture of the dried coagulate. Research has been carried out to reduce the undesirable flavour (Knorr et al., 1976; Wilhelm & Kempf, 1981), improve the texture (Meuser & Smolnik, 1979) and increase the solubility (Ney, 1979; Knorr, 1980), and this is continuing.

The most important limitations on the use of PPC in human food are toxins, especially trypsin inhibitors and glycoalkaloids (see Chapter 5). Present in potato fruit water, these are extracted along with PPC and hence become concentrated. Reductions of more than 40% in trypsin inhibitor activity, and of more than 80% in glycoalkaloid concentration, were achieved by heating protein concentrates or coagulates at 100 °C for 15 min before drying (Wojnowska et al., 1981). However, this thermal treatment considerably reduced the levels of amino acids in PPC and hence also reduced the chemical score and EAA index. Methods of eliminating, or of reducing to non-toxic levels, proteolytic enzyme inhibitors and glycoalkaloids, without adversely affecting the nutritional value and physicochemical properties of PPC, have still to be found. It has been suggested (Meuser & Smolnik, 1979) that protein products be further purified to produce protein isolates neutral in taste and smell. However, levels of toxic factors should still be carefully investigated.

Eriksen (1981) has reported on the nutritional value of a protein-rich fraction produced by air classification (separation) of spray-dried potato granules into starch- and protein-rich fractions. In this process, starch and undenatured protein concentrates are produced from potatoes without the addition of process water and hence without production of waste effluents. The chemical score and EAA index were marginally better for the protein-rich fraction than for the original granules. The EAA index of a protein isolate prepared from the protein-rich fraction was 96, comparable with that of egg protein. Such protein fractions could have great potential in food formulations. Indeed, other authors have emphasized the probable value of potato proteins as emulsifying agents in the food-processing industry (Finley & Hautala, 1976; Bakel, 1976; Holm & Eriksen, 1980; Wojnowska et al., 1981).

The use of PPC to replace part of the wheat flour in bread has been reviewed by Knorr (1979). His conclusion was that up to approximately 10% of wheat flour could be replaced by PPC without changing the volume of the bread. However, there is as yet no information on the nutritional quality of bread containing PPC.

The possibility of utilizing potato protein concentrates or isolates as major food ingredients or supplements, or even as a protein source in fabricated foods is still being explored. However, at least for the foreseeable future, such a possibility seems likely to be of more benefit to consumers in developed countries than it will be to those in developing countries, where the greatest need is for more protein.

References

Abrahamson, L. (1978). Food for infants and children in developing and industrialized countries. Ph.D. thesis. Institute of Nutrition, University of Uppsala.

Abouzied, M. M. & Mostafa, M. M. (1984). Production of single cell protein on potato-processing-waste media. *Proc. Nutr. Soc.* **43**: 93A.

Bakel, J. T. (1976). Potato protein for food uses: isolation and functional properties. Ph.D. thesis, University of Minnesota.

Bender, A. E. (1982). Evaluation of protein quality: methodological considerations. *Proc. Nutr. Soc.* **41**: 267–76.

Boody, G. & Desborough, S. (1984). *In vitro* digestibility and calculated PER as rapid methods for the nutritional evaluation of potato protein. *Qual. Plant. Plant Foods Hum. Nutr.* **34**: 27–39.

Burton, W. G. (1966). *The potato*, 2nd edn. Drukkerij Veenman BV Wageningen.

Chang, Y.-O. & Avery, E. E. (1969). Nutritive value of potato vs. rice protein. *J. Am. Diet. Assoc.* **55**: 565–7.

Chick, H. (1950). Supplementary nutritive value between the proteins of some common foods. *J. Am. Med. Womens Assoc.* **5**: 435–40.

Chick, H. & Slack, E. B. (1949). Distribution and nutritive value of the nitrogenous substances in the potato. *Biochem. J.* **45**: 211–21.

Dambois, I., Deeves, R. & Forwalter, J. (1978). Turns 3000 ppm food processing waste into byproduct, cuts sewage charges. *Food Processing* **39**: 156–7.

Davies, A. M. C. (1977). The free amino acids of tubers of potato varieties grown in England and Ireland. *Potato Res.* **20**: 9–21.

De Noord, K. G. (1976). Recovery of protein in potato starch manufacture. In W. T. Koetsier (ed.), *Chemical engineering in a changing world*. Elsevier Sci. Pub. Co., Amsterdam.

Desborough, S. L. & Lauer, F. (1977). Improvement of potato protein. II. Selection for protein and yield. *Am. Potato J.* **54**: 371–6.

Desborough, S. L., Liener, I. E. & Lulai, E. C. (1981). The nutritional quality of potato protein from intraspecific hybrids. *Qual. Plant. Plant Foods Hum. Nutr.* **31**: 11–20.

Desborough, S. L. & Weiser, C. J. (1974). Improving potato protein. I. Evaluation of selection techniques. *Am. Potato J.* **51**: 185–96.

Dreher, M. L., Dreher, C. J. & Berry, J. W. (1984). Starch digestibility of foods: a nutritional perspective. *CRC Crit. Rev. Food Sci. Nutr.* **20**: 47–51.

Dreher, M. L., Scheerens, J. C., Weber, C. W. & Berry, J. W. (1981). Nutritional evaluation of buffalo gourd root starch. *Nutr. Rep. Internat.* **23**: 1–8.

Eggum, B. O., Bach Knudsen, K. E. & Jacobsen, I. (1981). The effect of amino acid imbalance on nitrogen retention (biological value) in rats. *Br. J. Nutr.* **45**, 175–81.

Eppendorfer, W. H., Eggum, B. O. & Bille, S. W. (1979). Nutritive value of potato crude protein as influenced by manuring and amino acid composition. *J. Sci. Food Agric.* **30**: 361–8.

Eriksen, S. (1981). Protein nutritional quality of air-classified potato fractions. *J. Food Sci.* **46**: 540–2.

Espinola, C. N. Y. (1979). [Chemical analysis and evaluation of the protein quality of some potato varieties.] In Spanish. M.Sc. thesis, Universidad Nacional Agraria La Molina, Lima.

Fairweather-Tait, S. J. (1983). Studies on the availability of iron in potatoes. *Br. J. Nutr.* **50**: 15–23.

Finley, J. W. & Hautala, E. (1976). Recovery of soluble proteins from waste streams. *Food Product Dev.* **10**: 92–3.

Heisler, E. G., Siciliano, J., Treadway, R. H. & Woodward, C. F. (1959). Recovery of free amino acid compounds from potato starch processing water by use of ion exchange. *Am. Potato J.* **36**: 1–11.

Herrera, H. (1979). Potato protein: nutritional evaluation and utilization. Ph.D. thesis, Michigan State University.

Hoff, J. E., Jones, C. M., Wilcox, G. E. & Castro, M. D. (1971). The effect of nitrogen fertilization on the composition of the free amino acid pool of potato tubers. *Am. Potato J.* **48**: 390–4.

Hoff, J. E., Lam, S. L. & Erickson, H. T. (1978). Breeding for high protein and dry matter in the potato at Purdue University. *Purdue Univ. Agric. Exp. Station Res. Bull.* **953**.

Holm, F. & Eriksen, S. (1980). Emulsifying properties of undenatured potato protein concentrate. *J. Food Technol.* **15**: 71–83.

Hunnius, W., Fritz, A. & Munzert, M. (1976). [The effect of year of planting and weather on the protein contents of potatoes.] In German. *Landwirtsch. Forsch.* **29**: 141–8.

Hutterer, J. (1978). [Waste water utilization and removal at the Gmünd potato starch plant in Austria.] In German. *Starch/Stärke* **30**: 56–61.

Joseph, A. A., Roy Choudhuri, R. N., Indiramma, K., Narayana Rao, M., Swaminathan, M., Sreenivasan, A. & Subrahmanyan, V. (1963). Amino acid composition and nutritive value of the proteins of different varieties of potato. *Food Sci. (Mysore)* **12**: 255–7.

Joseph, K., Narayana Rao, M., Swaminathan, M. & Subrahmanyan, V. (1960). Effect of partial replacement of rice in poor rice diet by potato on the nutritive value of the diets. *Food Sci. (Mysore)* **9**: 41–3.

Kaldy, M. S. (1971). Evaluation of potato protein by amino acid analysis and dyebinding. Ph.D. thesis, Michigan State University.

Kaldy, M. S. & Markakis, P. (1972). Amino acid composition of selected potato varieties. *J. Food Sci.* **37**: 375–7.

Kapoor, A. C., Desborough, S. L. & Li, P. H. (1975). Extraction of non-protein nitrogen from potato tuber and its amino acid composition. *Potato Res.* **18**: 582–7.

Kaur, B. & Gupta, S. K. (1982). Utilization of potato for weaning food manufacture. *J. Food Sci. Technol.* **19**: 23–5.

Kies, C. and Fox, H. M. (1972). Effect of amino acid supplementation of dehydrated potatoes on protein nutritive value for human adults. *J. Food Sci.* **37**: 378–380.

Klein, L. B., Chandra, S. & Mondy, N. I. (1980). The effect of phosphorus fertilization on the chemical quality of Katahdin potatoes. *Am. Potato J.* **57**: 259–66.

Knorr, D. (1977). Protein recovery from waste effluents of potato processing plants. *J. Food Technol.* **12**: 563–80.

Knorr, D. (1978). Protein quality of the potato and potato protein concentrates. *Lebensm.-Wiss. Technol.* **11**: 109–15.

Knorr, D. (1979). Fortification of bread with potato products. *Starch/Stärke* **31**: 242–6.

Knorr, D. (1980). Effect of recovery methods on yield, quality and functional properties of potato protein concentrates. *J. Food Sci.* **45**: 1183–6.

Knorr, D., Höss, W. & Klaushofer, H. (1976). [Reduction of odour and taste of a commercial dried protein product recovered from potato vegetable water by heat coagulation.] In German. *Confructa* **21**: 166–180.

Kofrányi, E. (1973). Evaluation of traditional hypotheses on the biological value of proteins. *Nutr. Rep. Int.* **7**: 45–50.

Kofrányi, E., Jekat, F. & Müller-Wecker, H. (1970). The determination of the biological value of dietary proteins. XVI. The minimum protein requirement of humans, tested with mixtures of whole egg plus potato and maize plus beans. *Hoppe-Seyler's Z. Physiol. Chemie* **351**: 1485–93.

Kon, S. K. & Klein, A. (1928). The value of whole potato in human nutrition. *Biochem. J.* **22**: 258–60.

Labib, A. I. (1962). Potato proteins; their properties and nutritive value. Ph.D. thesis, Agricultural University of Wageningen.

Li, P. H. & Sayre, K. D. (1975). The protein, non-protein and total nitrogen in *Solanum tuberosum* ssp. *andigena* potatoes. *Am. Potato J.* **52**: 341–50.

López de Romaña, G., Graham, G. G., Madrid, S. & MacLean, W. C. (1981*a*). Prolonged consumption of potato based diets by infants and small children. *J. Nutr.* **111**: 1430–6.

López de Romaña, G., Graham, G. G., David Mellits, E. & MacLean, W. C. (1980). Utilization of the protein and energy of the white potato by human infants. *J. Nutr.* **110**: 1849–57.

López de Romaña, G., MacLean, W. C., Placko, R. P. & Graham, G. G. (1981*b*). Fasting and postprandial plasma free amino acids of infants and children consuming exclusively potato protein. *J. Nutr.* **111**: 1766–71.

Luescher, R. (1972). Genetic variability of 'available' methionine, total protein, specific gravity and other traits in tetraploid potatoes. Ph.D. thesis, Michigan State University.

McCollum, E. V., Simmonds, N. & Parsons, H. T. (1918). The dietary properties of the potato. *J. Biol. Chem.* **36**: 197–210.

McCollum, E. V., Simmonds, N. & Parsons, H. T. (1921). Supplementary protein values in foods. III. The supplementary dietary relations between the proteins of the cereal grains and the potato. *J. Biol. Chem.* **47**: 175–206.

Markakis, P. (1975). The nutritive quality of potato protein. In M. Friedman (ed.), *Protein nutritional quality of foods and feeds*, Part 2, *Quality factors – plant breeding, composition, processing and anti-nutrients*. Marcel Dekker, New York.

Meister, E. (1977). Genetic improvement of yield and nutritive value of tetraploid potatoes. Ph.D. thesis, Michigan State University.

Meister, E. & Thompson, N. R. (1976*a*). Physical-chemical methods for the recovery of protein from waste effluent of potato chip processing. *J. Agric. Food Chem.* **24**: 919–23.

Meister, E. & Thompson, N. R. (1976*b*). Protein quality of precipitate from

References

waste effluent of potato chip processing measured by biological methods. *J. Agric. Food Chem.* **24**: 924–6.

Meuser, F. & Smolnik, H.-D. (1979). Potato protein for human food use. *J. Am. Oil Chemists Soc.* **56**: 449–50.

Miedema, P., van Gelder, W. M. J. & Post, J. (1976). Coagulable protein in potato: screening method and prospects for breeding. *Euphytica* **25**: 663–70.

Millard, P. (1986). The nitrogen content of potato (*Solanum tuberosum* L.) tubers in relation to nitrogen application – the effect of amino acid composition and yields. *J. Sci. Food Agric.* **37**: 107–14.

Mulder, E. G. & Bakema, K. (1956). Effect of the nitrogen, phosphorus, potassium and magnesium nutrition of potato plants on the content of free amino acids and on the amino acid composition of the protein of the tubers. *Plant Soil* **7**: 135–66.

Neuberger, A. & Sanger, F. (1942). The nitrogen of the potato. *Biochem. J.* **36**: 662–71.

Ney, K. H. (1979). Taste of potato protein and its derivatives. *J. Am. Oil Chemists Soc.* **56**: 295–7.

Paul, A. A. & Southgate, D. A. T. (1978). *McCance and Widdowson's The Composition of Foods*, 4th edn, MRC Special Report no. 297. HMSO, London.

Peare, R. M. & Thompson, N. R. (1975). The influence of environmental and cultural practices on protein quality. In *EAPR Abstr. Conf. Papers*, 6th Triennial conference of the European Association for Potato Research, Wageningen.

Ponnampalam, R. & Mondy, N. I. (1983). Effect of baking and frying on nutritive value of potatoes. Nitrogenous constituents. *J. Food Sci.* **48**: 1613–16.

Rexen, B. (1976). Studies of protein of potatoes. *Potato Res.* **19**: 189–202.

Rose, M. S. & Cooper, L. F. (1917). The biological efficiency of potato nitrogen. *J. Biol. Chem.* **30**: 201–4.

Roy Choudhuri, R. N., Joseph, A. A., Sreenivas, H., Paul Jayaraj, A., Indiramma, K., Narayana Rao, M., Swaminathan, M., Sreenivasan, A. & Subrahmanyan, V. (1963). Nutritive value of poor Indian diets based on potato. *Food Sci. (Mysore)* **12**: 258–62.

Skogman, H. (1976). Production of symba-yeast from potato wastes. In *Food from waste*. AB Sorigona, Staffanstorp.

Smith, O. (1968). Chemical composition of the potato. In O. Smith (ed.), *Potatoes: production, storing, processing*. AVI Publishing Co. Inc., Westport, CT.

Snyder, J. C. & Desborough, S. L. (1980). Total protein and protein fractions in tubers of Group Andigena and *Phureja–Tuberosum* hybrids. *Qual. Plant. Plant Foods Hum. Nutr.* **30**: 123–34.

Strolle, E. O., Cording, J. & Aceto, N. C. (1973). Recovering potato proteins coagulated by injection heating. *J. Agric. Food Chem.* **21**: 974–7.

Strolle, E. O., Aceto, N. C., Stabile, R. L. & Turkot, V. A. (1980). Recovering useful byproducts from potato starch factory waste effluents – a feasibility study. *Food Technol.* **34**: 90–5.

Synge, R. L. M. (1977). Free amino acids of potato tubers: a survey of published results set out according to potato variety. *Potato Res.* **20**: 1–7.

Talley, E. A. (1983). Protein nutritive value of potatoes is improved by fertilization with nitrogen. *Am. Potato J.* **60**: 35–9.

Talley, E. A., Toma, R. B. & Orr, P. H. (1984). Amino acid composition of freshly harvested and stored potatoes. *Am. Potato J.* **61**: 267–79.

Thompson, N. R. (1977). The potato as a source of protein for a man. Plant protein production course lecture, Michigan State University.

Thompson, J. F. & Steward, F. C. (1952). The analysis of the alcohol-insoluble nitrogen of plants by quantitative procedures based on chromatography. II. The composition of the alcohol-soluble and insoluble fractions of the potato tuber. *J. Exp. Bot.* **3**: 170, 181–7.

Tikhonov, N. I. & Bychkov, V. A. (1969). [Contents of essential amino acids in potato tubers depending on the level of mineral nutrition.] In Russian. *Khim. Selsk. Khoz.* **7**: 5–7.

Tjørnholm, T., Baerug, R. & Roer, L. (1975). The amino acid composition of potato tubers as influenced by fertilizer supply, growing conditions and variety. In *EAPR Abstr. of Conf. Papers*, 6th Triennial conference of the European Association for Potato Research, Wageningen.

van Gelder, W. J. T. (1981). Conversion factor from nitrogen to protein for potato tuber protein. *Potato Res.* **24**: 423–5.

van Gelder, W. J. T. & Vonk, C. R. (1980). Amino acid composition of coagulable protein from tubers of 34 potato varieties and its relationship with protein content. *Potato Res.* **23**: 427–34.

Vigue, G. T. (1973). The influence of biochemical and environmental factors on protein synthesis and accumulation in potato. Ph.D. thesis, University of Minnesota.

Vigue, J. T. & Li, P. H. (1975). Correlation between methods to determine protein content of potato tubers. *HortScience* **10**: 625–7.

Watt, B. K. & Merrill, A. L. (1975). *Composition of foods: raw, processed, prepared*, Agriculture Handbook no. 8. U.S. Department of Agriculture, Washington, DC.

Wilhelm, E. & Kempf, W. (1981). [Progress in separation and stabilization of potato protein.] In German. *Starch/Stärke* **33**: 338–42.

Wojnowska, I., Poznanski, S. & Bednarski, W. (1981). Processing of potato protein concentrates and their properties. *J. Food Sci.* **47**: 167–72.

WHO (1973). *Energy and protein requirements*, Report of a joint FAO/WHO Ad Hoc Expert Committee, WHO Tech. Rep. Ser. 522. WHO, Geneva.

WHO (1985). *Energy and protein requirements*, Report of a joint FAO/WHO/UNU Expert Consultation, WHO Tech. Rep. Ser. 724. WHO, Geneva.

4

Effects of storage, cooking and processing on the nutritive value of potatoes

The nutritional value of the potato was considered in Chapter 2, where brief mention was made of the changes in nutrient content of raw potato that result from the various fates of the tubers after harvesting. Goddard & Matthews (1979) have stressed the need for data on the nutrient content of food in the form in which it is actually consumed so that planners may correctly assess intake of nutrients in order to provide a balanced diet. The purpose of this chapter, therefore, is to review current literature pertaining to the changes taking place in potato nutrient content as a result of storage, cooking or processing.

Not all of these changes are adverse or even very significant. Nutrient losses, however, do occur to a varying extent depending on the operation involved. It should be remembered, however, that post-harvest handling of some kind is often essential; the potato has to be cooked before consumption, and storage and processing are frequently needed to prevent seasonal gluts and to increase the availability of potatoes to consumers throughout the year. Some nutritional losses are therefore inevitable. The major points and extents of loss are given below as guides for workers in the fields of storage, nutrition, dietetics, catering and processing and to indicate possibilities for prevention or reduction of such losses.

Other vegetables also undergo adverse nutritional changes after harvesting. However, the potato's skin acts as a barrier, preventing or reducing leaching of nutrients into the cooking water. The skin itself is a source of some nutrients and may be consumed. In contrast, many other vegetables lack a protective skin and are subject to leaching losses during cooking. Some vegetables have inedible skins or skins that must be removed before cooking.

It is noticeable that gaps remain in our knowledge of nutritional changes, especially of those that occur during storage and processing. Some products lack sufficient systematic studies of the changes that take place during individual unit operations. Moreover, results from different authors are sometimes conflicting. Their inclusion in the review is intended to suggest indirectly the areas of interest which need further study or confirmation. Workers' results, no doubt, vary because of differences in the variety of potato and methods of storage, preparation, cooking or processing, as well as sampling and analytical procedures.

The freshly-harvested tuber is the point of reference used in this chapter to compare potatoes which have been stored, cooked or processed. It may appear that, in general, processed products have lower nutrient contents than home-cooked items. It should be borne in mind, however, that processors sometimes have the advantage of using freshly harvested tubers, and are also able to add back nutrients (e.g. ascorbic acid) lost during processing. Processed products may therefore have concentrations of some nutrients higher than those of previously stored potatoes badly prepared in the home. Furthermore, the laboratory preparations discussed here may not exactly simulate home-cooking procedures, and may only indicate what changes are likely to happen domestically. Baking of potatoes, where mentioned, implies oven-baking of tubers in their skins.

Decreases in moisture content and solid constituents during post-harvest operations may concentrate the retained nutrients so that losses may sometimes be of little importance if different potato preparations are compared on an equal-serving-weight basis. It must be assumed that, although some preparations, e.g. french fries or roast potatoes, may have higher energy contents than boiled potatoes, consumers tend to eat them all in equal quantities. The tables provided here and in Chapter 2, may be used to compare nutrient contents of equal servings of cooked and processed products. Occasionally losses or retentions of nutrients have been recalculated from the original data.

In the developing world, the nutritional effects of home-cooking procedures are currently of more importance than those of large-scale processing, which is limited in scale. However, the effects of processing have been included because of their importance in the developed world, and also because of possible future expansion of potato processing elsewhere. Part of the processing section is also devoted to the traditional methods of potato processing in the Andean highlands of Peru and Bolivia.

No attempt has been made to give details of storage or processing methods, these being outside the scope of this publication. Parts 1 and 3

Part 1: Storage

Storage conditions

Continuous production of potatoes throughout the year is virtually impossible in most countries. To increase potato availability between harvests and avoid large fluctuations in supply and therefore price, storage is required (Figure 4.1). Potato supplies are needed for domestic and institutional consumption, food industry processing, provision of seed, or industrial uses such as starch production. Nutritional changes during storage are only relevant to the use of the potato for

Figure 4.1. Potatoes destined for consumers are often stored at home and carried to local markets in small quantities to augment weekly household income. To market (left) in Kinigi, Rwanda, and (right) in Kunming, China.

feeding people. Detailed descriptions of post-harvest behaviour and principles of potato storage are outside the scope of this review, but are available elsewhere (Burton, 1978; Booth & Shaw, 1981; Rastovski *et al.*, 1981).

Good storage should maintain tubers in their most edible and marketable condition by preventing large moisture losses, spoilage by pathogens, attack by insects and animals, and sprout growth. It should also prevent large accumulation of sugars, which leads to an unpleasant sweet taste, and, particularly in fried processed products, to a dark coloration. Prevention of tuber greening and glycoalkaloid accumulation is also important.

During tuber growth, changes in nutrient content occur. The levels of various nutrients in the tuber at harvest depend not only upon variety and the cultural and environmental conditions under which the potatoes were grown but also upon maturity at time of harvest and the extent of damage which may have occurred during lifting and handling. At harvest, tubers are dormant; the axillary buds of the scale leaves (tuber 'eyes') are not actively growing. The dormancy period varies in length, depending mainly upon the variety of potato and storage temperature. At the end of dormancy, sprouts grow from the axillary buds. During dormancy and sprouting, further changes in nutrient content take place which affect the nutritional value of stored potatoes: these changes are less pronounced during dormancy, so storage conditions should aim to prolong the period and delay sprouting.

Normally, over a range of 4 °C to 21 °C, the lower the storage temperature the longer is the dormant period. Sprout growth is slow at temperatures of 5 °C and below. Above 5 °C, increasing temperature causes increased sprout growth up to about 20 °C; at even higher temperatures the growth rate decreases. However, lowering the storage temperature to below 10 °C causes an increase in sugar content which becomes marked below 6 °C. This decreases the culinary acceptability of tubers and increases the brown coloration of heat-processed products which occurs as a result of reactions between amino acids and reducing sugars. Tubers affected by low temperature sweetening can be 'reconditioned' (i.e. desweetened) by being held for about two weeks at 15 °C to 20 °C. However, sugar may accumulate after prolonged storage at higher temperatures – an occurrence known as senescent sweetening, which is irreversible.

Methods of storage vary from delayed harvesting, or storage in simple piles or 'clamps', to storage in buildings specially designed for the purpose, with controlled temperature and humidity. In the latter case,

Storage

potatoes can be maintained at a temperature suitable for retaining the quality characteristics appropriate to the purpose for which they are subsequently required. Before storage, tubers are generally allowed a period of about two weeks at 10 °C to 15 °C to undergo wound healing. According to Burton (1978), potatoes to be used for domestic consumption should then be stored at about 5 °C to avoid serious sprout growth and senescent sweetening. Low temperature sweetening should take place at a level which can be subsequently reduced during distribution. Tubers for later use in the food processing industry may be maintained at about 10 °C, which avoids disease and excessive sprout growth, but prevents a high accumulation of reducing sugars. Tubers stored for several months might undergo senescent sweetening at 10 °C and may be better stored at 7 °C to 8 °C, after wound healing. Any low temperature sweetening which occurs can be reversed by reconditioning prior to processing.

The use of different storage temperatures has been described briefly here to explain why authors have determined nutrient changes at many different temperatures. Most research has been done using low temperatures and in some cases controlled humidities. Information is not available about nutrient losses during simple on-farm storage relevant to developing countries, where temperatures and humidities are uncontrolled (Figure 4.2). Under these conditions temperatures are often high and humidities rather low most of the time. More information is needed in this field.

During storage tubers may lose moisture as a result of evaporation. Such losses will be at a minimum where relative humidity of the potato store is controlled, but will vary where humidity is uncontrolled. All concentrations of nutrients determined during storage should therefore be expressed either on a dry weight basis or adjusted for moisture loss. Strict accuracy in determinations of nutrient changes during storage should also include an adjustment for dry matter loss. This has been given at 10 °C as about 1.2% during the first month and 0.8% per month, thereafter rising to 1.5% per month when sprouting is well advanced (Burton, 1966). These values are lower at 5 °C, but higher at temperatures below or above 5 °C. Unless losses in dry matter (DM) are taken into account, nutrient retentions could be overestimated. Although most authors have considered moisture losses, DM losses have been ignored.

Recent years have seen extensive research into a new method of food preservation: irradiation. This is likely to be an important future development, although its acceptance has been delayed by fear and controversy, food irradiation having been mistakenly associated with the effects of ionizing radiation on living things, and wrongly classified as a food

Figure 4.2. Traditional method of storing potatoes in Rwanda (left) and in Peru (right).

Storage

additive. It is extremely useful for suppressing potato sprouting before storage, thus eliminating losses due to sprouting. Its economic feasibility for potato preservation is still open to question. However, it has been suggested that it might be economically attractive in warm, tropical countries where it could be used in combination with cool storage at 10 °C to 15 °C, rather than conventional cold storage at 2 °C to 4 °C. There is only one industrial potato irradiation plant operating at present, in Japan. Thomas (1984) has reviewed all aspects of the irradiation preservation of potatoes.

Changes in potato nutrients as a result of storage
Carbohydrates

Changes in carbohydrate chemistry during storage were described briefly in Chapter 1. The conversion of starch or sucrose to reducing sugars is generally of more interest because of its influence on flavour and colour in cooked or processed potatoes rather than its effect on nutritional value. In some processes, however, marked reductions in amino acid contents take place as a result of reactions between amino acids and carbohydrates during non-enzymic (Maillard) browning, particularly if the levels of reducing sugars are elevated as a result of storage. Care is usually taken to minimize carbohydrate changes by the use of appropriate storage temperatures or by the use of a period of reconditioning after storage. During traditional storage at higher temperatures, low temperature sweetening will not occur, but prolonged periods of such storage may raise the tuber sugar content as a result of senescent sweetening.

Nitrogenous constituents

Changes in nitrogenous constituents of potato tubers during storage have been studied either to detect changes in cell metabolism and growth processes affecting dormancy and sprout growth, or to identify free amino acid changes resulting from Maillard browning. From a nutritional viewpoint, little is known about qualitative or quantitative changes in potato N that occur during either traditional or commercial storage.

Total nitrogen

On the whole, changes in total N are small. Those reported by some authors (Yamaguchi *et al.*, 1960; Fitzpatrick & Porter, 1966) are not of a significant level. Weaver *et al.* (1978), however, reported a decrease ($P < 0.05$) of 8% in total N after four months of storage at 7 °C. Storage at the same temperature for only two months had no effect on total N. Toma *et al.* (1978*b*) showed a small increase in total N during eight months of cold storage, but this change was not significant in all the varieties studied. Nor was there any difference in this respect between storage temperatures of 3 °C or 7 °C. Talley *et al.* (1984), after holding two varieties for four months at 3 °C or 7 °C, found that N increased significantly at the lower storage temperature but not at the higher. Reconditioning at the end of storage has been found to increase total N only slightly (Toma *et al.*, 1978*b*) or to have no effect (Weaver *et al.*, 1978). No major changes in protein contents were observed during nine months of storage (at 10 °C for two varieties and at 7 °C for one other), but total and true protein increased slightly (Mazza *et al.*, 1983). No differences in total N (DWB) were found between tubers stored for five months at 20 °C or those in cold storage at 3 °C (Fernandez & Aguirre, 1975). Nor was there a difference between these and tubers stored at 20 °C and treated with chemicals or irradiation to prevent sprouting. Where authors have reported changes in total N, no attempt has been made to explain how these changes occurred.

Protein/non-protein nitrogen

Although the total N of stored potatoes has generally been reported to change very little, there is evidence of changes in individual nitrogenous constituents, but these have been studied mainly in the non-protein nitrogen (NPN) fraction and their nutritional significance is doubtful. An average loss of only about 3% protein was found for 12 samples stored at 3 °C to 4 °C between October and January (Desborough & Weiser, 1974).

Some evidence indicates that protein breakdown may occur as a result of sprouting. Burton (1978) pointed out that the potential for this is limited if the tuber is to continue as a living entity during sprouting because tuber protein is enzymic, and not storage, in nature. Klein *et al.* (1982) observed a decrease in the ratio protein N : NPN in the cortex of three varieties and in the pith of two of these when potatoes sprouted. This suggested an increase in free amino acids at the expense of protein in order to make amino acids available for translocation to the sprout tissues. However, although sprouts contained much higher levels of nitrogenous constituents than the tuber tissues, there were not significant reductions in these constituents in the tuber. Tagawa and Okazawa (cited by Burton, 1978) showed an increase in soluble N in the terminal bud and cortex at the end of dormancy and a corresponding decrease in the protein of the pith. Sirenko (cited by Porter & Heinze, 1965) found a loss of protein in four varieties at the end of eight months of cold storage and indicated that most of the loss could be accounted for by an increase in NPN. Lower storage temperature apparently decreased this transformation. Míča (1978a) found that protein N decreased with length of storage at both 2 °C and 10 °C, although changes in total N were small. In a further study (Míča, 1978b), free amino acid content was higher at the end of storage (May) than at the beginning (November), there being an initial decrease and then an increase in free amino acids. These changes were more marked at 10 °C than at 2 °C. Fernandez & Aguirre (1975) found that tubers at 3 °C had a higher percentage of free amino acids after five months of storage than those stored at 20 °C, even though the latter had sprouted. However, there is no indication in this work of the changes taking place during the actual storage period.

Habib & Brown (1957) reported little or no change in free amino acid composition of four cultivars stored at about 4 °C, but reconditioning at 23 °C caused a marked decrease in total free amino acids and complete loss of arginine, histidine and lysine. Fitzpatrick & Porter (1966), however, found a general increase in all free amino acids after cold storage and reconditioning. They attributed this to metabolic degradation of the protein occurring as tubers sprouted during the later stages of reconditioning. Other workers have recorded decreases in NPN (Mondy & Rieley, 1964) or free amino acid content (Weaver et al., 1978) during cold storage.

Only slight changes were found in most quantitatively important free amino acids in potatoes stored for eight months at 3 °C, followed by one month at 7 °C or at 10 °C (Talley et al., 1964). These changes did not show any consistent pattern in relation to storage. One exception was proline, which increased, especially during the last three months. The late change in proline was confirmed (Talley & Porter, 1970) but other amino acid levels altered randomly. Sweeney et al. (1969), in contrast, found increases in all free amino acids, except glutamic acid, during the first two months of storage at 13 °C. Between two and five months at this temperature, all free amino acids except leucine decreased quantitatively, but apart from aspartic and glutamic acids and lysine they were still higher than in the initial unstored potatoes. During one month of storage at 21 °C, all except aspartic and glutamic acids showed increases. Amino acid changes (increases or decreases) were generally greater at 21 °C than at 13 °C.

Changes in the individual total amino acids have been studied in two varieties by Talley et al. (1984) for a four-month storage period at 3 °C or at 7 °C. Changes were either small or insignificant. There was no change in methionine, isoleucine or tyrosine at either temperature. Lysine, tryptophan, aspartic acid and arginine increased at both temperatures. Threonine, valine, leucine, phenylalanine and histidine increased at 3 °C, but not at 7 °C, whereas cystine increased at 7 °C but not at 3 °C. Míča (1978c) observed a fall in the lysine content of six out of ten varieties during the first months of storage either at 2 °C or at 10 °C, and then a slight increase. In two of these varieties the decrease continued, although at a reduced rate. In only one variety was the lysine content higher at the end of storage than initially. Eight of ten varieties studied had higher lysine contents after six months of storage at 2 °C than at 10 °C. Thomas (1984), reviewing the effects of irradiation on the nitrogenous constituents, noted that one group of workers found a large increase in lysine content during short-term (up to one month) storage of irradiated potatoes. It is not known whether the higher levels are sustained during more prolonged storage.

It is difficult to draw any firm conclusions from the literature about the nature of changes in nitrogen constituents during storage and the possible nutritional significance. Different workers have used different temperatures and periods of storage, reconditioning has been carried out in some cases and sprouting has sometimes occurred. The changes reported have not been assessed from a nutritional viewpoint but are probably not important in this respect.

Fibre

Appleman & Miller (1926) reported an increase in crude fibre of immature potatoes during storage, but no change in that of mature potatoes. Little variation occurred in crude fibre levels in several North American varieties stored at 3.3 °C for up to eight months (Toma *et al.*, 1978*b*). As pointed out in Chapter 2, crude fibre determinations have largely been discontinued, as they measure only a small and variable part of the dietary fibre. There are no reports at present of changes in potato dietary fibre during storage. It is unlikely that this would alter quantitatively. Changes in the physiological properties of plant dietary fibres at different stages of development and ripening are known to take place, but the nutritional effects of such changes in the potato have not yet been investigated.

Vitamins
Ascorbic acid

A knowledge of post-harvest changes in ascorbic acid is important where potatoes form a staple part of the diet and provide a large proportion of the daily intake of this vitamin. The changes due to storage (reviewed by Burton, 1966, 1978) may result from synthesis, conversion to dehydroascorbic acid (nutritionally useful), diketogulonic acid (not useful), or other reactions leading to loss. The net result is a significant loss of ascorbic acid during storage and this has been well documented. The literature, however, deals mostly with storage under conditions of controlled temperature and humidity: there is only one recent study (Linnemann et al., 1985) of ascorbic acid losses under conditions simulating traditional farm storage in the developing world, where temperatures may be fairly high, and where potatoes may undergo moisture loss and sprouting. More information should be sought in this area, since variations in storage temperature and the occurrence of sprouting can affect the final concentration of ascorbic acid.

Thomas (1984) reported that, according to most workers, ascorbic acid is stable during and after irradiation. Ascorbic acid levels decline during subsequent early storage periods, but after prolonged storage are comparable with, or even greater than, those of unirradiated tubers stored under identical conditions.

Various authors have investigated the extent of ascorbic acid loss during cold storage. Decreases have ranged from about 40% to 60% over the course of several months (Sweeney et al., 1969; Augustin et al., 1978a; Faulks et al., 1982). Shekhar et al. (1978), however, found a loss of about 50% of the initial reduced form of ascorbic acid in tubers stored at 5.5 °C for only four weeks. 'White Rose' potatoes held for 30 weeks at 5 °C or at 10 °C (Yamaguchi et al., 1960), lost 72% and 78%, respectively, of their total ascorbic acid (reduced form + dehydroascorbic acid). Where reduced ascorbic acid only has been determined, it is possible that losses have been overestimated, as part of the ascorbic acid could have been converted into dehydroascorbic acid. Roine et al. (1955), for example, observed a disappearance of 49% of ascorbic acid after 8.5 months of storage. Part of this had been converted to dehydroascorbic acid and the actual vitamin loss was 40%. An earlier study by Smith & Gillies (quoted by Leichsenring et al., 1951) had found dehydroascorbic acid levels to be low in freshly harvested tubers but to increase rapidly during storage and to amount to one-third of the total ascorbic acid after six months. Similarly, Wills et al. (1984) found no dehydroascorbic acid in fresh potatoes, but after six weeks of storage 36% of the total ascorbic acid were as dehydroascorbate. Leichsenring et al. (1957) found losses in dehydroascorbic and diketogulonic acids as well as in ascorbic acid during storage, and hence could not account for losses in the nutritionally active forms by conversion into diketogulonic acid. It is conceivable that oxidation was proceeding beyond the formation of diketogulonic acid.

Losses have been found to take place most rapidly during the early part of storage. One-third to one-half of the total loss recorded over seven months of storage took place during the first month (Murphy, 1946). Losses of total ascorbic acid from 'Chippewa' and 'Triumph' varieties held at 1 °C occurred continuously throughout storage, but were most rapid during the first six weeks (Leichsenring et al., 1957). The sharpest decrease in total ascorbic acid content occurred during the first three to four months of storage at 7 °C (Augustin et al., 1978a), at 12 °C (Mareschi et al., 1983) or at an unspecified temperature (Finglas & Faulks, 1984), followed by a levelling out or less pronounced decrease during the following months, depending on variety.

Furthermore, at any storage temperature, initial differences in tuber ascorbic acid levels immediately after harvest are reduced, so that at the end of storage there may be comparatively little difference in ascorbic acid contents. Although varietal or seasonal differences in initial ascorbic acid values were not completely nullified after seven months of storage, they were greatly diminished (Murphy, 1946; Roine et al., 1955). Augustin et al. (1978a) found that, after eight months of storage, total ascorbic acid values were similar whether they had been initially high or low immediately after harvest. Moreover they were close to the final total value (DWB) of 'White Rose' potatoes stored for 30 weeks (Yamaguchi et al., 1960).

Storage temperature may also influence the final content of ascorbic acid. All but one of the early sources quoted by Murphy (1946) found ascorbic acid values to be higher in potatoes stored at temperatures above 10 °C than in those stored below 10 °C. It is likely that these authors only determined reduced ascorbic acid. Murphy (1946) determined that the optimum temperature for maximum retention of reduced ascorbic acid and maintenance of good physical quality is 10 °C over a period of seven months of storage, although at 15 °C, 18 °C or 21 °C more vitamin C was retained than at 0 °C or 2 °C. Of the three storage temperatures studied by Effmert et al. (1961), smallest decrease in reduced ascorbic acid during seven months of storage of 20 varieties was at 15 °C, followed by 1 °C and then 5 °C. The least loss of reduced ascorbic acid occurred between 10 °C and 21 °C according to Werner & Leverton (1946). As the temperature was lowered from 10 °C to 4.5 °C the loss steadily increased. There is evidence that the biosynthesis of ascorbic acid is related to carbohydrate metabolism. As the latter is considerably altered at low temperatures (viz. low temperature sweetening), it is perhaps not surprising that ascorbic acid is also affected.

Linnemann et al. (1985) studied the effects of high storage temperatures on the reduced ascorbic acid content of 'Bintje' potatoes, in an attempt to simulate conditions in developing countries. Tubers were stored at 16 °C or at 28 °C and at 55% to 60% rel. hum., for 12 weeks. At the end of this period the 16 °C tubers had a few small sprouts, whereas the 28 °C tubers were soft, shrivelled and had long, thick-set sprouts. The low initial ascorbic acid content of 8.2 mg/100 g (FWB) had increased to 10.1 and 10.5 mg/100 g (FWB) in the 16 °C and 28 °C stored tubers, respectively. Although these increases are lower if tuber weight losses occurring during storage are taken into account, they compare with a slight decrease which took place in tubers stored at 7 °C (95% to 98% rel. hum.). The authors attributed the increases at the higher temperatures to the occurrence of sprouting and increased respiration.

These findings indicate that vitamin C losses from potatoes in traditional stores in developing countries are likely to be lower than those during low temperature storage. Further study is required to determine changes in total active ascorbic acid forms at different storage temperatures and relative humidities.

The net changes in tuber ascorbic acid as a result of sprouting depend upon synthesis in the tuber and translocation to the sprouts, which varies according to the vigour of sprout growth. In general this results in a decrease in tuber vitamin C during the early stages of sprout growth, followed by a temporary increase and by another decrease (Burton, 1978). Bantan *et al.* (1977) found a higher ascorbic acid content in the sprouts than in the rest of the tuber after eight months of storage. Transfer of ascorbic acid from tuber to sprouts lowers the tuber nutritional value (as sprouts are removed before cooking) and sprouting should therefore be prevented.

Reports of the effect of reconditioning on ascorbic acid levels after low temperature storage are conflicting. Augustin *et al.* (1978*a*) reported no increase in ascorbic acid on reconditioning (conditions of which were not stated). However, Leichsenring *et al.* (1957) found that potatoes held at 24 °C for three weeks following storage at 1 °C had higher ascorbic acid contents (retention 65% to 67%) than those held continuously at 1 °C (55% to 59% retention), without reconditioning.

The B group vitamins

Reports of changes in thiamin content during storage are variable; in general, changes appear to be small and rather erratic during the storage period. Only Meiklejohn (1943) found a large loss in thiamin during storage, but attributed this to sprouting rather than storage *per se*, as the sprouts had a higher thiamin content than the tubers from which they had grown. After six to seven months of storage about 30% to 50%

of the vitamin was lost from the tuber by translocation to the sprouts. At temperatures of 5 °C or 10 °C, Yamaguchi *et al.* (1960) found no significant change in thiamin in 'White Rose' potatoes, even after 30 weeks of storage, but did not mention whether sprouting had occurred. Other authors reported slight but significant increases in some varieties of North American potatoes. Leichsenring *et al.* (1951) noted a slight increase in thiamin over 24 weeks of storage (conditions not mentioned), even on a dry weight basis, although the overall trend for one of the four varieties ('Chippewa') studied showed a slight decrease. A small but significant increase in thiamin was found after eight months of storage of two varieties stored at 3 °C and in most of the samples of six varieties stored at 7 °C (Augustin *et al.*, 1978*a*). However, at the latter temperature, the changes in varieties 'Norchip' and 'Kennebec' were not significant. No information is available on thiamin changes in potatoes stored for short periods at ambient temperatures. On the whole, cold storage appears to have little effect on thiamin values even after prolonged periods.

The overall insignificant change in riboflavin found in 'White Rose' potatoes after 30 weeks of storage at either 5 °C or 10 °C (Yamaguchi *et al.*, 1960) and the small and erratic changes in riboflavin in six varieties after eight months of storage at 7 °C (Augustin *et al.*, 1978*a*) are of little dietary importance. With freshly harvested potato, 100 g can provide only 2% of the RDA of riboflavin.

The overall change in niacin was not significant for 30 weeks of storage at 5 °C or at 10 °C in 'White Rose' (Yamaguchi *et al.*, 1960). The figures for 5 °C, however, show a slight increase up to six weeks and a subsequent decline to a value a little lower than that of the freshly harvested tubers. Similarly, in a study by Page & Hanning (1963), varieties 'Cobbler' and 'Triumph' showed significant increases of 36% and 19%, respectively, in niacin concentration, which reached their peak after one or two months. This was followed by a decline to values approximating to those of the freshly harvested tubers. The authors indicated that the increases during the first one to two months of storage were greater than could be accounted for by tuber weight losses (although these were assumed, rather than measured). They also reported that varieties 'Cobbler' or 'Sebago' stored at 24 °C for four weeks had lower niacin levels than when stored at 4.5 °C for a similar period, although the extent of the decrease was not reported. A general decrease in niacin values after eight months of cold storage at 7 °C has been reported (Augustin *et al.*, 1978*a*). It is not possible to know whether this decline followed an earlier increase, as reported in other papers, as potatoes were not analysed until four months of storage time had elapsed. In general, therefore, prolonged periods of cold storage have little effect on niacin values. Short periods may lead to tubers with values higher than those of freshly harvested potatoes. Storage for short periods at ambient temperatures, however, may cause a decrease in niacin.

Increases in pyridoxine during storage have been demonstrated by the two groups. There was a continuous increase in pyridoxine in both 'Cobbler' and 'Triumph' varieties from harvest time to the end of a six-month storage period at 4.5 °C (Page & Hanning, 1963). The total increases were 152% and 86%, respectively, for the two varieties. Moreover, in 'Cobbler' and 'Sebago' stored at approximately 24 °C, pyridoxine increased at a more rapid rate than at 4.5 °C, the difference being greater in 'Cobbler' than in 'Sebago'. Augustin *et al.* (1978*a*) also found significant increases in pyridoxine during storage; for example, variety 'Katahdin' increased by 126% at the end of eight months of storage at 7 °C. A re-examination of their data revealed similar increases in the other five varieties and in the two varieties stored at 3 °C. The authors point out that it is not known if this increase is due to synthesis of

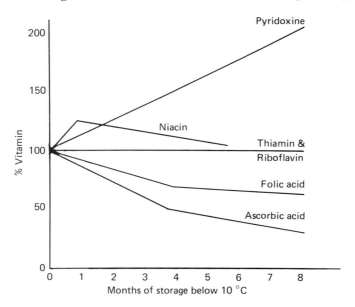

Figure 4.3. Trends in vitamin concentration changes during cold storage.

the vitamin or its release from a bound form during the early stages of storage.

In contrast to the large increase in pyridoxine, total folic acid decreased significantly in all six varieties analysed over the eight-month cold storage period (Augustin et al., 1978a); losses ranged from 17% to 40% (DWB). The figures show that the losses were most rapid during the first four months and thereafter showed little change up to eight months.

Figure 4.3 shows the general trends which occur in vitamin concentration changes during cold storage at temperatures below 10 °C.

There has been little work on the effect of irradiation on tuber B group vitamins. Thomas (1984), however, mentions two groups of workers who found no changes in contents of B vitamins in irradiated potatoes.

Minerals

The sparse information available indicates that changes in mineral constituents during storage are insignificant. In general there was no change in the total ash content of two varieties at 3 °C or of six varieties at 7 °C over a period of four to eight months of storage (Toma et al., 1978a). Losses of minerals on cooking after 24 weeks of storage at 6 °C were not greater than those after only two weeks of storage (Faulks et al., 1982). Therefore, apparently storage alone had not altered mineral composition. Míča (1979) found for 10 cultivars that, although phosphorus contents tended to increase during storage and boiling and potassium decreased, these changes were not significant. Yamaguchi et al. (1960) observed no significant changes in the contents of calcium, iron or phosphorus in 'White Rose' potatoes held at 5 °C or 10 °C for 30 weeks.

Comments on nutritional changes during storage

In general, few adverse nutritional changes occur in stored potatoes, with the exception of losses of ascorbic and folic acids, both of which suffer further losses as a result of cooking or processing. Storage produces a beneficial increase in the concentration of pyridoxine.

Previously stored potatoes should preferably be steamed or boiled in their skins when prepared domestically (see Part 2), to maximize vitamin retentions. Assuming a 50% loss of ascorbic acid as a result of four to five months of storage and a further 20% loss by boiling the tubers unpeeled, 100 g of stored, cooked potatoes can still supply about 25% of the RDA for vitamin C. These potatoes would also supply about 4% of the RDA for folic acid, similar quantities of niacin and thiamin as the freshly harvested tubers (about 8% to 10% of the RDA) and about 20% of the pyridoxine RDA.

Effects of storage, cooking and processing

Part 2: Domestic preparation

Peeling

Apart from removal of dirt by washing, one of the first steps in the preparation of potatoes, either domestically or on a large scale may be peeling (Figure 4.4). This is removal of the outer layers of the potato; however, the parts removed may vary from the periderm only to both periderm and much of the cortex, depending upon the type, shape and age of the tuber, presence or absence of damaged parts, and method of peeling. Waste and weight loss will vary accordingly. Removal of peel to a uniform depth of 1.5 mm would remove nearly 20% by weight from a

Figure 4.4. Children in Bisate, Rwanda, peel potatoes for the family's evening meal while their mother is away cultivating the family fields.

50 g tuber varying to about 10% from a 200 g tuber (Burton, 1974). A decrease in weight loss by peeling with increasing tuber size (which results in a decreased ratio of surface area:volume) was also shown by Weaver et al. (1979). The same authors showed (with a method of peeling in caustic soda) that for tubers of equal weight, percentage weight loss due to peel removal increases as the potato becomes more flat or elongated, because of an increase in the ratio surface area:tuber volume.

The method of domestic peeling can greatly affect quantity of loss through peeling. Scraping young potatoes removed about 5% of weight as waste, whilst peeling increased this to 20% to 24% (Szkilladziowa et al., 1977). The average weight loss on peeling and trimming maincrop (old) potatoes was 12.6% and scraping early (new) potatoes was 8% in one season for four British potato varieties (Finglas & Faulks, 1984). The most important factor leading to loss was damage in the form of penetrating cracks, followed by superficial crushing and bruising. Careful domestic peeling may remove 10% (by weight) of the potato, but often as much as 25% may be removed (Burton, 1974). Zobel (1979) cites an author who noted losses of up to 50% by peeling of variety 'Ora'. Augustin et al. (1979c) found that peel fractions of the whole tuber, when peel was removed with a domestic peeling knife either before or after cooking, were 6%, 2% and 10% in the case of peeled raw potatoes, those peeled after boiling, and those peeled after baking, respectively. To minimize waste, potatoes should be peeled after and not before boiling, when cooked domestically, unless damage is extensive.

Nutritional aspects

There are two effects to be considered: (1) The loss of nutrients from the peel itself when peel is removed before or after cooking and not consumed. This depends upon the distribution of nutrients within the tuber. (2) The fact that the peel forms a barrier preventing loss of nutrients during cooking. This point is discussed more fully in the section on boiling peeled *versus* unpeeled potatoes.

Distribution of nutrients in the tuber

Augustin et al. (1979c) and Talley et al. (1983) investigated the truth of the commonly held consumer belief that potato nutrients are concentrated in the peel. They concluded that this belief is not justified, although total N, crude fibre, ash, riboflavin and folic acid (Augustin et al., 1979c) and amino acids (Talley et al., 1983) occurred as higher percentages of the raw tuber peel than of the flesh when 6% of the tuber was removed as peel. The percentages of tuber peel and flesh occurring as ascorbic acid, niacin, and pyridoxine were approximately equal, whereas thiamin occurred as a higher proportion of the flesh than of the peel. The resulting contents of nutrients in the peel as percentages of those in the whole tuber are presented in Table 4.1.

The distribution of nutrients in the tuber has also been shown by Zobel (1979); the concentration of minerals in the outer layers of the tuber is particularly noticeable; this has been confirmed for some minerals and trace elements (Johnston et al., 1968; Bretzloff, 1971; Kubisk et al., 1978; Mondy & Ponnampalam, 1983). Peeling before cooking, therefore, should result in a loss of minerals; Maga (personal communication), for example, found a 31% reduction in both zinc and calcium contents when mature potatoes were peeled.

According to Zobel (1979), removal of approximately 20% of the fresh weight by peeling removes only 10% of the ascorbic acid, which is in a lower concentration in the outer part of the potato, but if 40% of the potato is removed by peeling, 45% of the ascorbic acid is lost. He quotes a study by Potapov, who found, using a domestic peeler, 6.9% to 36% loss of ascorbic acid, depending upon season and cultivars.

Table 4.1. *Nutrient content of peel as a percentage of that of the whole tuber*[a]

Constituent	Raw[b]	Boiled[c]	Baked[d]
Dry matter	4.7	1.9	17.7
Total N	8.3	2.9	16.7
Crude fibre	34.4	14.9	37.3
Ash	15.9	3.3	17.0
Ascorbic acid	5.0	1.1	10.5
Thiamin	1.7	0.6	7.8
Riboflavin	9.4	2.9	28.6
Niacin	4.1	1.8	15.0
Folic acid	8.0	1.6	15.2
Pyridoxine	5.6	1.7	15.6

[a] From Augustin et al. (1979c). Reprinted from *J. Food Sci.* 1979, **44** (3): 806. Copyright © by the Institute of Food Technologists.
[b] Peeled with domestic peeler before cooking; peel fraction of the whole tuber was 6%.
[c] Boiled in water and then peeled with a household knife; peel fraction of the whole tuber was 2%.
[d] Oven-baked and then peeled with a household knife; peel fraction of the whole tuber was 10%.

The tuber centre has been suggested to be the site of highest thiamin concentration (Meiklejohn, 1943; Augustin *et al.*, 1979c). So peeling should not result in significant losses of thiamin.

The outer parts of the potato contain mainly insoluble protein, of negligible nutritional value (see Chapter 3), and removal of the skin by careful peeling does not adversely affect the protein quality of the potato. Rats given diets containing steamed, peeled potato as the sole source of N grew better than those on a diet with steamed, whole potato (Chick & Slack, 1949).

Total N, crude fibre, ash, riboflavin (Augustin *et al.*, 1979c), and amino acids (Talley *et al.*, 1983) occurred as higher proportions of the peel, removed from potatoes after boiling, than of the boiled flesh. The quantities of nutrients in peel removed from boiled potatoes expressed as percentages of total tuber nutrients were lower in all cases than those of raw peel (see Table 4.1). This must have been partly because the peel itself represented a small proportion (only 2%) of the whole tuber, and also because some heat destruction or leaching of nutrients into the cooking water had occurred. This again suggests that potatoes should be peeled after boiling.

In the case of baked potatoes (Table 4.1), total N, crude fibre, ash, riboflavin, niacin, folic acid and pyridoxine were all proportionately higher in the peel than in the flesh, and all peel nutrients were present as higher percentages of the total nutrients than they were in the peel of raw or boiled potatoes. This is undoubtedly due to the loss of moisture from the tuber during baking and a consequent concentration of the peel nutrients, so whether a baked potato is consumed with the skin should be considered when evaluating its nutritional value.

Boiling unpeeled versus peeled potatoes

In about 1795, one Count Rumford wrote that 'it seems to be the unanimous opinion of those who are acquainted with these useful vegetables that the best way of cooking them is to boil them simply, and with their skins on, in water' (Anon., 1875). Boiling is the most common method of preparing potatoes domestically in every part of the world because it is also the cheapest. Boiling will therefore be considered separately from other domestic cooking methods.

Without exception, experimental work carried out to compare the nutrient content of potatoes boiled in their skins and then peeled with that of potatoes peeled before boiling shows the nutritional advantages of the former method. It is noteworthy that potatoes are usually boiled intact by consumers in some developing countries, in contrast to the normal practice of peeling tubers before cooking in e.g. the USA and Britain. Cutting tubers exposes a greater surface area to the water and increases losses of nutrients by leaching.

Moisture

Insignificant changes in moisture content have been reported for both boiled, peeled or boiled, unpeeled potatoes (Hughes, 1958; Toma *et al.*, 1978*a*). Herrera (1979) reported a decrease in weight after boiling and a corresponding increase in total solids, which seemed to indicate that weight loss was due to water loss. Only about 1% to 2% weight loss occurred in potatoes boiled with an intact skin, but increased when potatoes were peeled or cut. The greatest differences were between uncut potatoes with and without skin. Once either of these had been cut into halves or quarters there was little difference between peeled and unpeeled samples. Weaver *et al.* (1983) found a 2% increase in water content when peeled potatoes were boiled and a 9% decrease in content of solids as a result of leaching from the tissue into the cooking water.

Carbohydrate

The major part of potato carbohydrate is present as starch. The digestibility of cooked and uncooked starches from various foods including potato has been reviewed by Dreher *et al.* (1984), who placed potato starch in the group of least digestible food starches. There have been various experiments in which raw potato starch was fed to humans and caused symptoms such as violent stomach cramps (McCay *et al.*, 1975), and such preparations cause caecal hyperotrophy and death in rats (El-Harith *et al.*, 1976). The latter effects were subsequently attributed to the resistance of potato starch to digestion by pancreatic amylase (Walker & El-Harith, 1978), and were lost when the starch was gelatinized.

Cooking either peeled or unpeeled potatoes increases the digestibility of potato starch. The results of a study *in vitro* with pancreatic amylase into the effects of cooking potatoes on starch digestibility (Hellendoorn *et al.*, 1970) are shown in Figure 4.5. Raw starch was barely digested; partly cooked starch from potatoes heated in water at 70 °C for 20 min and cooled immediately was incompletely digestible, and the digestibility of the starch increased with cooking time. Potatoes judged to be adequately cooked at 25 to 30 min had slightly lower digestibility than those 'overcooked' at 40 min. This work is applicable to adult consumers, but infants and small children have only low levels of amylase activity in the intestine. In addition, Englyst *et al.* (1982) and Jones *et al.* (1985) found varying quantities of starch that is resistant to enzymic hydrolysis *in vitro*, in samples of cooked potato.

Domestic preparation

When cooked potato was incorporated into the diets of infants (López de Romaña et al., 1980), consumption of increasing amounts of potato (representing 25%, 50% and 75% of the total dietary energy), was associated with increasing indigestibility of carbohydrate. Including potato in weaning diets at levels exceeding 50% of the total energy was suggested by these authors to be impractical. De Vizia et al. (1975), however, found that potato starch, whether given as biscuits cooked in water for 10 min or as cooked potato flour was almost completely absorbed by infants. Unfortunately, the manner of reporting the quantities of potato starch fed to the children in the two different studies is such as to make it impossible to compare them, and further clarification is needed to determine the maximum quantity of potatoes, properly cooked in the home, which can be tolerated by young children. This is especially important in view of the potato's low energy density, and the possible adverse effect of potato starch on protein digestibility (see Chapter 3, p. 73).

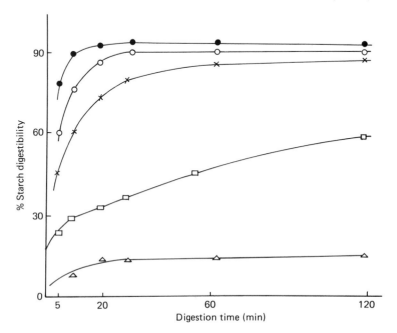

Figure 4.5. Digestibility *in vitro* of raw and cooked potato starch. Measurements were made with starch from potatoes: uncooked (△); heated in water at 70°C for 20 min and then cooled (□); boiled for 15 min (×), 25 min (○) or 40 min (●). (From Hellendoorn et al., 1970.)

Nitrogenous constituents

In general negligible or small changes in the content of nitrogenous constituents take place when potatoes are boiled whole, unpeeled (Roy Choudhuri et al., 1963b; Schwerdtfeger, 1969; Toma et al., 1978a; Talley et al., 1983). In a study by Herrera (1979), the loss of total N in boiled, unpeeled potatoes was only 0.8%. This rose to more than 4% when unpeeled potatoes were halved or quartered, and was highest in boiled, peeled potatoes – 6.5% in the whole tuber and 10% in halved or quartered tubers. Predominant amino acids in the water from potatoes boiled with intact skins were aspartic and glutamic acids. Eppendorfer et al. (1979) also found that the 3% to 4% of tuber N found in the cooking water after boiling potatoes for 25 min occurred mainly as aspartic acid (asparagine) and glutamic acid (glutamine). This is not surprising because soluble NPN is more easily leached into the cooking water than is protein N. The average losses of total N in four varieties of maincrop (old), and new, peeled, boiled potatoes can be calculated from the figures of Finglas & Faulks (1984) to be 8% and 26%, respectively; 13% and 31%, respectively, of the NPN were lost.

Aspartic and glutamic acids also predominated in the boiled water from halved or quartered potatoes (Herrera, 1979), but there was an increase in the boiled water content of isoleucine, lysine and histidine in quartered as compared to intact or halved potatoes. Losses of methionine and cystine to the cooking water progressively increased from intact to halved to quartered potatoes, the losses of methionine being much greater than those of cystine. This was probably because more methionine than cystine is present in the free form. Herrera therefore advised boiling potatoes whole with intact skins.

Losses of amino acids from peeled and cut potatoes were greater than from those unpeeled (Schwerdtfeger, 1969), but they were small from both types of preparation. Significant losses (12% to 20%) of N were reported for two out of four varieties when uncut, peeled tubers were boiled, whereas there was no loss from the unpeeled potatoes (Toma et al., 1978a). Desborough & Weiser (1974) reported a 50% loss of protein for 12 genotypes after halved tubers were boiled for 30 min.

When potatoes are peeled and/or cut and stored sometime before cooking, they readily undergo enzymic browning, involving chlorogenic acid or amino acids. A study of the effect of enzymic browning on the amino acids of the potato (Davies & Laird, 1976) showed that substantial losses of the sulphur-containing amino acids took place during browning in all four varieties investigated. Such losses could affect the nutritional value of potato N, which is limiting in the sulphur amino acids. Avoidance of amino acid losses is achieved easily by cooking peeled potatoes immediately or, for example, immersing tubers in water to prevent enzymic browning. The same study found no change in the availability of lysine by enzymic browning. Boiling of 'Russet Burbank' tubers decreased available lysine by about 11% (Herrera, 1979).

Fibre

There is a variety of changes in dietary fibre on boiling reported in the literature. This is probably due in part to varying definitions of dietary fibre and to differing methods of determination (see Chapter 2). Paul & Southgate's (1978) tables show a decrease in quantity of dietary fibre in the flesh of potatoes boiled for 30 min as compared with raw potatoes. No change in the dietary fibre content (DWB) of the flesh of maincrop (old) and early (new) potatoes, peeled and boiled, can be shown from the average figures of Finglas & Faulks (1984) for four varieties. Johnston & Oliver (1982) found increases in detergent-extracted fibre for both boiled, peeled and boiled, unpeeled potatoes. They suggested that such apparent increases were probably the result of an interaction between food components to produce compounds (other than lignin, which remained unchanged) insoluble under the conditions of digestion used for the fibre determinations (see also Englyst *et al.*, 1982; Jones *et al.*, 1985).

Higher levels of total dietary fibre in several varieties of whole boiled potatoes than in those boiled without the skin have been found (Jones *et al.*, 1985). Determinations of the hexose, pentose, uronic acid, cellulose and lignin levels in these samples indicated qualitative differences, the nutritional significance of which has yet to be determined.

Vitamins

Vitamins are susceptible to losses during cooking or processing caused by (a) leaching into cooking or blanching water, (b) destruction by heat treatment, (c) chemical changes such as oxidation. A vitamin may be affected by one or all of these changes, depending on its stability to changes in pH, oxidation, light and heat. Table 4.2 shows the responses of the major vitamins found in potatoes to these conditions. It can be seen that the vitamins least stable to all factors are ascorbic and folic acids and the most stable vitamin is niacin.

Ascorbic acid

Some 40 to 50 years ago, it was established that the smallest losses of ascorbic acid occur when unpeeled potatoes are boiled or steamed. When 'Chippewa' potatoes were cooked by various methods, the losses were significant, except when they were boiled in their skins (Leichsenring *et al.*, 1951). The mean total ascorbic acid retention for two Indian varieties of whole, unpeeled tubers cooked in boiling water was 84% (Swaminathan & Gangwar, 1961), whereas only 62% was retained in boiled, peeled potatoes. Only 10% and 21%, respectively, of the losses were due to leaching into the cooking water. The rest must have been destroyed by heat. The retentions are similar to those which can be calculated from the figures for reduced ascorbic acid of Hentschel (1969), i.e. 83% for boiled, unpeeled and 73% for boiled, peeled, cut potatoes, and to the average 60% retention of total ascorbic acid for four varieties of peeled, boiled tubers found by Finglas & Faulks (1984). Ascorbic acid retentions were greatest for whole, unpeeled, and least for peeled, quartered, boiled potatoes (Bucko et al., 1977). Whole, boiled, unpeeled potatoes retained 80% of their reduced ascorbic acid, whereas boiled, peeled tubers retained only 74% (Augustin et al., 1978b).

B group vitamins

Thiamin may be little affected by boiling. There was only a small difference between thiamin values of boiled, unpeeled or boiled, peeled, cut potatoes as compared to raw ones (Hentschel, 1969). Overall thiamin retentions were high (88%) for several varieties of whole boiled potatoes (Augustin et al., 1978b). Peeling did not alter retention; however, losses of thiamin ranging from 0 to 40% (average 20%) were noted in four varieties of peeled, boiled potatoes (Finglas & Faulks, 1984).

Hentschel (1969) found niacin retentions on boiling of 75% and 65%, respectively, in unpeeled and peeled potatoes. Other authors (Leichsenring et al., 1951; Augustin et al., 1978b) reported niacin retention of 100% for boiled unpeeled, and of 92% and about 80%, respectively, for peeled whole or peeled, quartered, boiled potatoes. Similarly, 82% niacin

Table 4.2. *Stability of major vitamins found in potatoes under various conditions*

Vitamin	pH 7	Acid	Alkali	Air	Light	Heat
Ascorbic acid	U	S	U	U	U	U
Thiamin	U	S	U	U	S	U
Riboflavin	S	S	U	S	U	U
Niacin	S	S	S	S	S	S
Pyridoxine	S	S	S	S	U	U
Folic acid	U	U	S	U	U	U

S = Stable; U = Unstable.
Information taken from R. S. Harris, in *Nutritional evaluation of food processing* (R. S. Harris & E. Karmas, eds.). Copyright 1975 by AVI Publishing Company, Westport, CT.

Domestic preparation

retention was reported in peeled, halved tubers (Page & Hanning, 1963), and an average of 85% (range 70% to 100%) in peeled whole tubers (Finglas & Faulks, 1984).

Pyridoxine was also shown to be unaffected when potatoes were boiled whole in their skins (Augustin *et al.*, 1978*b*). Mean retention values of 100% for unpeeled tubers were reduced to 85% of the original value when whole tubers were cooked peeled. These values are higher than those found by Page & Hanning (1963), who showed a mean pyridoxine retention of 80% in 58 samples of boiled, peeled tubers. However, these authors cut their tubers in half before boiling.

Riboflavin content was found to be unaltered in potatoes boiled in their skin (Hentschel, 1969) and only slightly lower in boiled peeled, cut potatoes. In comparison, the overall riboflavin retention for four North American varieties was 87% in whole, unpeeled, and only 75% in whole, peeled, boiled tubers (Augustin *et al.*, 1978*b*), and was, on average, 80% (range 60% to 100%) in peeled, boiled British varieties (Finglas & Faulks, 1984). Folic acid was decreased to 81% and 72% of the raw value in unpeeled, boiled and peeled, boiled tubers, respectively (Augustin *et al.*, 1978*b*), and to an average of 80% of the raw value in peeled, boiled tubers (range 60% to 90%) (Finglas & Faulks, 1984).

Figure 4.6, summarizes the approximate losses of vitamins that result from boiling potatoes either in their skins or previously peeled.

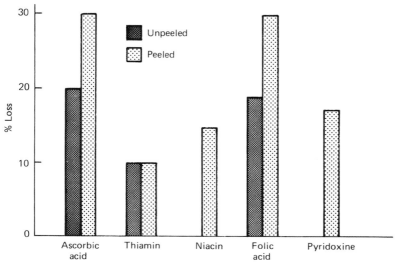

Figure 4.6. Vitamin losses in potatoes that have been boiled either peeled or unpeeled. The data are averages of findings by various authors mentioned in the text. There were no losses of niacin or pyridoxine in unpeeled potatoes.

Minerals

Research providing information on changes in mineral constituents during boiling is sparse.

Boiling and peeling resulted in a loss of 18% total ash in four varieties (Toma *et al.*, 1978*a*). There was no difference in the ash contents of raw or unpeeled, boiled potatoes (Leichsenring *et al.*, 1951; Roy Choudhuri *et al.*, 1963*b*; Toma *et al.*, 1978*a*).

For two varieties boiled with or without jackets in different types of containers, the changes in zinc, copper, manganese, nickel and chromium contents were small and nutritionally insignificant (Seiler *et al.*, 1977). A striking reduction in iron content took place in peeled or unpeeled potatoes when boiled in containers with a non-metallic surface. Potatoes boiled without skin lost more potassium and magnesium than potatoes with skin, but there was no loss in calcium in either case. There were no losses of calcium, phosphorus or iron when three Indian varieties were boiled in their skins (Roy Choudhuri *et al.*, 1983*b*). Three North American varieties, boiled whole with and without skins in a stainless steel pan and analysed for 14 minerals (True *et al.*, 1979), showed that retentions were generally high (over 90%). In one variety ('Norchip'), iron content was reduced to only 54% of its original value by boiling with or without skin. Decreases in calcium, magnesium, potassium, copper and phosphorus also were found for 'Norchip' without skin. Losses for potassium and copper, 10% to 15% for sodium, magnesium, phosphorus and iron and 0% for calcium and zinc were shown for boiled, peeled, mature maincrop (old) potatoes (Finglas & Faulks, 1984). They were assumed to result from leaching into the cooking water, as they occurred only on boiling. Reductions were generally greater in early (new) potatoes; 30% zinc, for example, was lost from the earlies, which, according to Finglas & Faulks (1984), might indicate changes in distribution and association of zinc with other cellular components during maturation.

Losses of minerals through boiling, except in the case of iron, are probably smaller than losses due to careless peeling before cooking.

Other domestic methods of potato preparation

Methods of domestic preparation of potatoes, other than boiling, may require one or more stages to achieve the end result. For example, potatoes baked in their skins require only one stage of preparation. However, potatoes may be soaked in water before cooking, fried potatoes may be boiled before frying, and the potato used for potato dumplings is grated or minced and at least part of it boiled twice before serving. The literature shows generally greater losses of nutrients from tubers subjected to more than one step of preparation compared to those more simply prepared. As cooking methods such as baking and frying involve decreases in moisture content of tubers, nutrients are concentrated, and true changes during cooking are impossible to assess when nutrient contents are expressed on a fresh weight basis. For this reason, authors normally compare nutrient contents of raw potatoes and those cooked by different methods on the basis of either dry weight or an equal quantity of DM.

Nitrogenous constituents

Changes in total N or individual amino acids have been found to be mostly small when potatoes are subjected to only one step of cooking. Losses are greater when the method of preparation is more complicated.

No differences were found in the content of total N of any of three Indian varieties either deep-fat fried from raw or baked (Roy Choudhuri et al., 1963b). There was a significant decrease in the nitrogen content of two of four North American varieties when baked (Toma et al., 1978a). The average (2.5%) was less than that caused by boiling peeled potatoes (8%).

Decreases in total N, protein N and individual amino acids were small when previously peeled potatoes were fried (Schwerdtfeger, 1969) although greater than those from boiled potatoes. The greatest losses occurred in potatoes peeled, cut and boiled before frying, and the least from potatoes fried from raw. Jaswal (1973) also found small losses of 5% and 7% for bound and free amino acids, respectively, presumed to be as a result of carbohydrate–amino acid interactions, when raw potatoes were french fried. Much greater losses (approximately 50%) in total N, protein N and individual amino acids were sustained in the case of potato dumplings (Schwerdtfeger, 1969). Protein N and NPN were reduced in the same proportions so that there was little change in the EAA index, which was similar to that of fried potato. It seems likely that soluble nitrogenous constituents were lost partly in the juice squeezed out during grating or mincing, and partly by leaching into the cooking water, particularly from the twice-boiled part of the potato.

In recent years, fried potato peels, prepared from the cortex tissue of baked potato (approximately 40% of the whole tuber), have become very popular in the USA but there is little information about their nutritional value. Ponnampalam & Mondy (1983) studied changes in the nitrogenous constituents in the baked or fried cortex and pith (which is used after removal of the cortex) in three varieties. Baking reduced total N by 5% to 18% and total amino acids by 5% in the cortex, whereas it increased total

N (3% to 20%), NPN (9% to 28%) and total amino acids (13%) in the pith. Conventional oven-baking of tubers at high temperatures may result in membrane damage and breakdown of some of the true protein into free amino acids. Nitrogenous constituents, especially free amino acids may migrate from the outer cortex to the inner pith tissues. Frying the baked cortex or raw pith resulted in significant losses of total N, NPN and amino acids. The baked and then fried cortex lost 29% to 43% total N, 20% to 35% NPN and 45% amino acids. Among the essential amino acids, the greatest loss was of lysine. The authors speculate that losses during frying could have resulted from generated volatile constituents, loss of nitrogenous constituents into the cooking oil and Maillard reactions between amino acids and carbohydrates. Overall, there were greater losses in nitrogenous constituents with frying than with baking. Hence the nutritional value of the outer potato tissues served as fried peels was undoubtedly less than that of those eaten baked.

Fibre

Information is lacking on the effects of various domestic cooking processes on dietary fibre. Paul & Southgate's (1978) tables show no change of dietary fibre content in the flesh of baked potatoes. There was little change in dietary fibre content of the flesh of baked, roasted or french-fried 'Sebago' variety potatoes on a dry weight basis (Jones *et al.*, 1985) when compared with raw potato. There was an apparent increase in the samples on an as-eaten basis, presumably due to concentration through moisture loss. Enzyme-resistant starch (determined separately) was found in all samples at a concentration higher than that in raw potato.

Vitamins
Ascorbic acid

As in determining ascorbic acid changes in stored potatoes, some workers have determined only the reduced form, others the total active ascorbic acid, in cooked potatoes. When conditions favour oxidation, for example during warm-holding, mixing or mashing, determinations of only the reduced form of the vitamin may overestimate losses.

Leichsenring *et al.* (1951) showed that allowing peeled, whole or quartered tubers to soak in tap water at room temperature for up to 3 h did not reduce ascorbic acid levels. Soaking hand-peeled potatoes in tap water for longer periods of 20 or 40 h produced significant decreases of 7% and 10%, respectively (Zarnegar & Bender, 1971). Similarly, soaking peeled, sliced tubers for one, two or three days brought about ascorbic acid decreases of 8%, 13%, and 28%, respectively (Hadziyev & Steele,

1976). The same authors produced an increase in ascorbic acid by aerating potato slices at 22 °C in a dark, moist atmosphere. Ascorbic acid had increased 180% after 48 h due to biosynthesis in the potato tissue. This finding, the authors felt, justified an old household practice of peeling and slicing potatoes the day before use and keeping them overnight wrapped in a moist cheese-cloth or towel. It has been pointed out (McCay *et al.*, 1975; Oguntona & Bender, 1976) that losses in tuber total ascorbic acid can take place during large-scale cooking or catering resulting from peeling and cutting potatoes a long time before they are to be cooked, soaking them in water and discarding the water.

Retention of reduced ascorbic acid was found to be the same (about 85%) whether peeled potatoes were boiled, steamed or pressure-cooked (Leichsenring *et al.*, 1951). However, if peeled, halved and boiled potatoes were subsequently stored in a refrigerator for 24 h, only 47% of the original vitamin content remained. Retention decreased even further, to only 18%, after refrigerator storage of 72 h. Leichsenring *et al.* (1957) later showed, however, that part of this loss was apparently due to conversion of ascorbic acid to dehydroascorbic acid. Hence, in 'Chippewa' and 'Triumph' varieties for example, losses of total active forms of vitamin C were only 48% and 21%, respectively, as opposed to losses of reduced ascorbic acid of 54% and 34%.

Boiled potatoes may be stored for some time, as indicated above, when required for subsequent preparations, potato salad for instance. When boiled, unpeeled potatoes were stored in the refrigerator (Hentschel, 1969), the reduced ascorbic acid content fell from 12.5 mg/100 g freshly cooked tubers to 8.3 mg/100 g after one night (55% retention of the original content of raw potatoes) and 7.5 mg/100 g after two nights (50% retention). When potato salad was made from the stored potatoes, reduced ascorbic acid fell to 5.3 mg/100 g (35% retention). The author attributed some of the loss to oxidation of ascorbic acid during cutting of the potatoes and intensive mixing during preparation of salad. He also found that potato dumplings retained only 14% of the reduced ascorbic acid found in the raw starting material, and explained this by the fact that preparation involves boiling the potato twice. The retentions of ascorbic acid found by this author are probably underestimated due to a failure to determine dehydroascorbic acid.

Streightoff *et al.* (1946) reported that, during large-scale preparation, only 11% of the total ascorbic acid in their peeled potatoes was lost on steaming, but they found 24% to 68% loss when the cooked potatoes were mashed, and 58% loss when the mashed potatoes were warm-held on a steam table for 45 min. Such destruction is largely due to oxidation.

In the preparation of mashed potato (reviewed by Priestley, 1979), significant destruction of ascorbic acid, ranging from 30% to 80%, may occur. However, Mareschi *et al.* (1983) found an average 40% loss of total ascorbic acid in both mashed and whole, peeled, boiled potatoes. Augustin and Demoura (Augustin *et al.*, 1978*b*) found reduced ascorbic acid loss of 60% when mashed, boiled potatoes were held at room temperature for 1 h.

Large losses of ascorbic acid also occur during hash-browning (i.e. boiling, shredding, and subsequent frying of the cooked shreds). Retentions of 57% of reduced and 30% of total original ascorbic acid have been reported in two cases of hash-browned potatoes (Leichsenring *et al.*, 1951; Pelletier *et al.*, 1977).

When Indian varieties were baked in a conventional oven, they lost between 50% and 56% of their ascorbic acid, compared to only 20% to 28% loss for potatoes boiled unpeeled (Roy Choudhuri *et al.*, 1963*b*). Augustin *et al.* (1978*b*), however, found an overall reduced ascorbate loss for two varieties of baked North American potatoes of only 25%, similar to that for boiled, peeled, but slightly higher than that of boiled, unpeeled potatoes. Streightoff *et al.* (1946) had previously reported a loss of 28% total ascorbic acid from baked, compared to only 13% from boiled, peeled potatoes. The difference between baked and boiled, unpeeled potatoes is probably due to the higher temperature employed for oven-baking. In general, oven-baking appears to cause slightly greater destruction of vitamin C than does boiling whole tubers in their skins.

It has also been shown (Leichsenring *et al.*, 1951) that holding baked potatoes after baking greatly reduces ascorbic acid content. Immature tubers analysed immediately after baking had retained 92% of their total ascorbic acid. Retention fell to only 78% after holding at room temperature for 4 h, and to only 35% when the baked tubers were held at 75 °C for 3 h.

Frying also reduces ascorbic acid content. Losses of 50% to 59% of total ascorbic acid were noted when sliced or shredded potatoes were washed and then deep- or shallow-fat fried (Swaminathan & Gangwar, 1961; Roy Choudhuri *et al.*, 1963*b*) and were greater than when tubers were boiled either peeled or unpeeled. However, only part of the total loss was due to frying *per se*. In the case of slices, about 26%, and in the case of shreds about 45%, of the total loss was due to washing before frying (Swaminathan & Gangwar, 1961). Sautéed potatoes (boiled for 15 min and fried for 7 min) lost 44% of their initial total ascorbic acid and small- (0.4 cm) and normal-sized (0.8 cm) french-fries lost 37% to 46% ascorbic acid, the least loss occurring in normal-sized fries (Mareschi *et al.*, 1983).

B-group vitamins

Thiamin is fairly heat-stable during the domestic preparation of potatoes. Baked tubers retained 86% of their thiamin, a retention similar to that of boiled tubers (Augustin *et al.*, 1978*b*). Frying potatoes raw, peeled or previously boiled unpeeled, produced no change in their thiamin content (Hentschel, 1969). However, frying peeled, cut, boiled tubers reduced thiamin by about 40%. The greatest reduction takes place if peeled potatoes are left soaking in water for long periods before subsequent cooking. Oguntona & Bender (1976) found that soaking peeled half-tubers in water for 16 h resulted in thiamin losses of over 40% in the outer layer of the potato and of 20% even at a depth of 10 mm inside the tuber. Soaking raw, chipped potatoes (i.e. prepared for subsequent french frying) in water for 16 h caused a loss of 40% of the original thiamin. Subsequent frying of either the half-tubers or french fries caused further losses (10% and 5%, respectively), but these were small compared to the losses by soaking.

Niacin, being heat stable, is mainly lost by leaching into cooking water, especially when tubers are peeled or cut before boiling. No apparent loss of niacin was observed in either baked, roast or french-fried potatoes by Finglas & Faulks (1984). Niacin was reduced by only 4% to 8% by frying raw potatoes (Hentschel, 1969). Boiling peeled, cut potatoes, followed by frying, reduced niacin to 61% of its original value, but only part of this loss (about 9%) was due to frying. However, frying did not reduce niacin in potatoes that had been previously boiled in their jackets. Losses of niacin by frying were generally small.

The overall retention of niacin in four baked varieties was 93%, similar to that of boiled, peeled and a little lower than that of boiled, unpeeled potatoes (Augustin *et al.*, 1978*b*). A similar high mean retention of 96% was found for 58 samples of baked potatoes by Page & Hanning (1963), and 100% retention by Leichsenring *et al.* (1951). Furthermore, the niacin level was not affected by mashing, holding or hash-browning previously boiled, unpeeled tubers. Most of the niacin lost from boiled, peeled tubers was present in the cooking water (Leichsenring *et al.*, 1951; Page & Hanning, 1963). Cooking peeled, quartered potatoes in a pressure cooker in a smaller volume of water than that used for normal boiling, resulted in significantly greater retentions of niacin than boiling similarly prepared tubers (Leichsenring *et al.*, 1951).

The mean retention of pyridoxine in baked potatoes has been found to be 91% (Page & Hanning, 1963; Augustin *et al.*, 1978*b*). This is higher than that of boiled, peeled potatoes, but somewhat lower than that of boiled, unpeeled potatoes. The leaching of niacin and pyridoxine into the cooking water was found to be similar in boiled potatoes (Page &

Hanning, 1963) but destruction by heat was somewhat greater for pyridoxine, than for niacin, being 8.8% and 4.2%, respectively, in baked potatoes.

Riboflavin is partially destroyed by heat during dry methods of cooking. The overall retention of riboflavin in four varieties of baked potatoes was 77%, similar to that of boiled, peeled tubers (Augustin et al., 1978b). Moreover, riboflavin was partially destroyed when potatoes were fried, the least losses (up to 20%) occurring in potatoes fried from raw and the greatest (55% to 65%) in potatoes peeled, cut, boiled and then fried (Hentschel, 1969). Unlike ascorbic acid, riboflavin content was not affected in the preparation of potato salad from potatoes previously boiled in their skins. This is not surprising as riboflavin is not affected by atmospheric oxygen.

Decreases of folic acid in cooked foods can result from both heat destruction and leaching into the cooking water. The overall retention of total folic acid for four varieties of baked potatoes was 71%, similar to that of boiled, peeled potatoes, two of the varieties showing low retentions of only 48% and 54% (Augustin et al., 1978b). This was the lowest retention of any of the vitamins determined by these authors in boiled and baked potatoes. However, there was no apparent loss of folic acid in potatoes as eaten, when baked, roast or french fried, according to Finglas & Faulks (1984), presumably through concentration due to water loss.

The percentage retentions of ascorbic acid, thiamin, riboflavin, niacin, folic acid and pyridoxine in micro-wave-cooked potatoes were found to be 73, 95, 87, 103, 88 and 96, respectively (Augustin et al., 1978b). In general, these are similar to those of boiled, unpeeled potatoes (Augustin et al., 1978b). This method of cooking is at present rarely used domestically outside the USA and some European countries. The approximate vitamin losses caused by various commonly used domestic methods of preparation are summarized in Table 4.6, and can be compared with those produced by various processing methods.

Minerals

The total ash content is unaffected by holding, mashing or hash-browning potatoes previously cooked whole, unpeeled (Leichsenring et al., 1951) or by oven-baking (Toma et al., 1978b).

Sodium, potassium, calcium, magnesium, phosphorus, iron, zinc, iodine, boron, copper, manganese, molybdenum and selenium were virtually unaffected by baking in three North American varieties (True et al., 1979), except in the case of 'Norchip', which showed unaccountably low retentions of calcium, copper and iron. Baked 'Russet Burbank' tubers contained less calcium and copper on an equal serving-weight basis than peeled boiled tubers, but greater quantities of magnesium, potassium, sodium, manganese, zinc and iron (Weaver et al., 1983). Conventional baking reduced the potassium, phosphorus and iron contents of the outer cortical tuber tissues by 10% to 13%, 4% to 12% and 19% to 31%, respectively, in three varieties, whilst the contents increased by 14% to 23%, 2% to 9% and 2% to 8%, respectively, in the inner pith tissues (Mondy & Ponnampalam, 1983). These minerals had apparently migrated from the outer to the inner tissues during baking. Subsequent frying of the baked cortex and raw pith (as in the production of fried potato peels for sale in the USA) significantly ($P < 0.05$) reduced the cortical and pith contents of iron, manganese, copper and zinc. These changes are as yet unexplained. Calcium, magnesium and phosphorus contents were not affected when previously boiled potatoes were subsequently fried (Hentschel, 1969). However, retentions of these three minerals were only 63%, 24% and 24%, respectively, in potato dumplings. The apparent increases in sodium, potassium, calcium, phosphorus, iron, copper and zinc found in baked, roasted or french-fried potatoes on an as-eaten basis were a result of water loss (Finglas & Faulks, 1984). There was no evidence of any loss of minerals as a result of these preparation methods.

Retentions of a variety of minerals and trace elements in three varieties during micro-wave cooking were generally high. Notable exceptions were unexplained losses of aluminium, copper and iron in 'Norchip' (True et al., 1979).

Comments on nutritional changes during domestic cooking

Domestic cooking methods cause some nutrient losses, although the extent of these depends upon the particular nutrient in question and the type of preparation. Nevertheless, concentration of the retained nutrients due to losses of moisture and solids during cooking can mean that nutrient contents in a given quantity of a potato preparation such as baked or french fried are similar to those in an equal quantity of raw potato or potato boiled in its skin. True comparisons of the nutritional value of potatoes prepared in different ways should therefore be made on an equal serving weight basis (see e.g. Table 2.8, p. 34–5). In general, nutrient contents on this basis are similar to those of the raw potato.

Energy values of roasted and french-fried potatoes are considerably higher than those of other preparations, due to absorption of fat during cooking. The levels of ascorbic and folic acids are lower in all preparations than in the raw potato, as they are reduced significantly by cooking losses.

When potatoes are to be boiled, they should be boiled whole in their skins and the skins removed after cooking, if so desired. Intact boiled potatoes contain greater quantities of nutrients than those peeled and cut before boiling.

Nutrient changes, apart from that of ascorbic acid, have been little studied in potato preparations such as potato dumplings, hash-browns and salads, which require several stages for completion. Levels of ascorbic acid in these preparations are low and may be almost negligible if potatoes have been stored previously for long periods.

Part 3: Processing

Potato processing has been practised in the highland areas of Peru and Bolivia for at least 2000 years. The dried products known as *chuño* and *papa seca* are a vital part of the diet in these areas, and are produced by methods unchanged over the years.

On an industrial scale, however, processing is confined mainly to developed countries. Some developing countries, for example in South America, process potatoes into snack foods or instant mashed potato, but the market for these products is currently small compared to that for fresh potatoes, due to economic restrictions. It is possible that future demand for such products and potatoes in the form of other convenience foods may increase. In some countries, for example India, processing of potatoes is desirable to avoid gluts and the consequent difficulty of storing large quantities of potatoes during periods of extremely high temperatures. Dehydration using solar energy has been suggested (Singh & Verma, 1979), as a means of coping with the problem, particularly as a cottage industry in rural areas (Nankar & Nankar, 1979).

Large-scale potato processing began in the USA. In 1940, about 2% of all potatoes consumed domestically were processed by frying, dehydration, freezing or canning, but by 1970 the corresponding figure was 51% (Hampson, 1972). The decline in consumption of fresh potatoes in the USA has been matched by a significant rise in that of processed products, so that overall consumption has not decreased. The change from fresh to processed potatoes in developed countries is due partly to enhanced efficiency in processing, which reduced the relatively high cost of canned, dried and frozen forms and partly to factors such as increased demand for convenience foods, fast foods and picnic snacks.

Processing

Potato products may be classified as follows:

(1) Fried non-snack products (e.g. frozen french fries (chips)).
(2) Non-fried frozen products (e.g. potato patties, mashed potato).
(3) Snack products (e.g. chips (crisps)).
(4) Dehydrated products (e.g. potato granules, flakes, powder).
(5) Preserves (e.g. canned potatoes).
(6) Other products (e.g. potato salad, pre-peeled potatoes).

These will be discussed in detail below.

There is also a host of smaller-scale potato products, and patents are appearing constantly for new ones. Some of these products have been described by Feustel (1975) and Smith (1977).

It is not the purpose of this review to describe methods of processing in detail, but rather the nutritional changes which have taken place as a result of the operations employed. It is hoped that this will serve as a source of reference in countries engaged in potato processing now or in the future. Flow diagrams of the major processes are included as a guide to the operations mentioned during descriptions of the nutritional effects of processing, although it should be noted that these operations may vary from one processing plant/company to another.

The traditional processes employed in the high Andes also are briefly described and the limited amount of information available on the nutritional changes involved are reviewed.

Large-scale processing
Pre-peeled potatoes

Production of pre-peeled potatoes (described by Feinberg *et al.*, 1975) is a growing industry in developed countries. 'Pre-peeled' refers to potatoes peeled, preserved from discoloration, and cold-stored. They are perishable and have only a relatively short shelf-life, but are supplied to restaurants, canteens and retail establishments, which therefore do not need to invest in their own peeling machine. Potatoes may be whole, or cut into strips for french frying.

Potatoes are prepared for almost any type of processing, including pre-peeling, by the operations shown in Figure 4.7. In the case of pre-peeled potatoes these are followed by cutting (if desired), immersion in a sulphite solution for a few minutes to prevent enzymic browning reactions, draining, packing and refrigeration. The two most important operations are peeling and sulphiting. The information reviewed below on the nutritional effects of peeling is applicable to any of the processes described later.

Figure 4.7. Pre-processing operations.

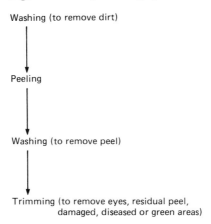

Peeling

Efficiency of peeling is necessary to minimize losses of raw material and production of waste which must be disposed of. Peeling losses vary with size and shape of tuber (Weaver *et al.*, 1979), depth of eyes and length of storage (Smith, 1975). The deeper the eyes or the longer the storage period the greater the loss.

Methods employed include abrasion-peeling, lye-peeling (immersion in sodium hydroxide solution at a low or high temperature), steam-peeling (use of steam under high pressure which is released rapidly after an adequate contact time) or a combination of lye- and steam-peeling.

The distribution of nutrients within the tuber, which affects the extent of nutrient losses by peeling, has been described above (see pp. 101–3). It is evident from this that any method of peeling will remove significant amounts of minerals, as there are proportionately greater amounts of them in the outer tuber layers. An abrasive peeling method that damages surface tissues of the tuber, producing a rough or spongy surface results in loss of juice containing soluble constituents including mineral salts (Zobel, 1979). The extent of mineral losses, however, has not been investigated in relation to large-scale peeling methods.

There is relatively little information about vitamin losses resulting from different potato peeling processes. Some information is summarized in Table 4.3. Peeling by steam retains the greatest quantities of ascorbic acid, thiamin, riboflavin and niacin. However, the figures indicate that even the lowest losses of thiamin and riboflavin are substantial. Zobel (1979) reports that losses of ascorbic acid from the surface zone, denatured by lye-peeling, are less than 5% and that he could find no

Table 4.3. *Vitamin losses as a result of different methods of peeling*

Method of peeling	Vitamin losses (% of fresh unpeeled tuber)			
	Ascorbic acid	Thiamin	Riboflavin	Niacin
Abrasion	$10.5^{a,b}$	—	—	—
Lye-steam combination	6.5^a	$32–35^c$	$25–26^c$	$10–23^c$
Steam	3.0^a	$18–20^c$	$15–16^c$	$5–5.5^c$

[a] From Zobel (1979).
[b] No significant loss of ascorbic acid was reported for mechanically peeled potatoes by Zarnegar & Bender (1971). It is conceivable that losses vary with type of peeler used.
[c] From Gorun (1978a).

significant difference in ascorbic acid content of seven cultivars peeled with lye or with a domestic peeler. However, Alarcón (1977) reported a loss of about 19% of ascorbic acid by lye-peeling. This is also considerably higher than the lye-steam combination reported in Table 4.3. Probably the extent of vitamin losses from lye-peeling depends upon such factors as time of immersion of tubers in lye solution and temperature and concentration of the solution. More information is required about effects of different methods of peeling on nutritive value, including those on pyridoxine and folic acid, which have not been investigated.

Sulphiting

Pre-peeled potatoes are immersed in a solution (usually of sodium metabisulphite) for a few minutes to inhibit discoloration by enzymic browning. They may then be stored refrigerated for several days before cooking and consumption.

As sulphite is a reducing agent, it should have little effect on ascorbic acid. Mudambi & Hanning (1962) found only a slight reduction in ascorbic acid due to sulphiting and suggested that this was due to leaching of the vitamin into the sulphite solution. In various texts on potato processing, mention is made of the addition of sulphite (apart from its other functions) as a means of protecting vitamin C from heat destruction.

Thiamin is converted by sulphite into inactive thiazole and pyramidine sulphonic acid (see Figure 4.8). Losses of thiamin through sulphiting of pre-peeled potatoes are considerable, depending upon the amount of surface area of potato exposed to sulphite, the concentration of solution, quantity of sulphite absorbed and length of subsequent storage time.

Figure 4.8. Destruction of thiamin by sulphiting.

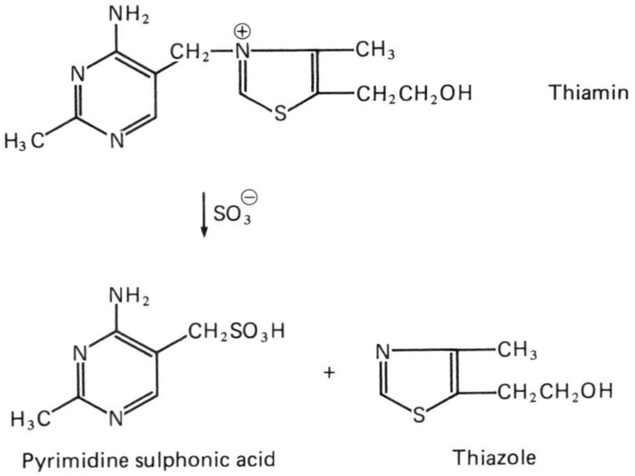

[Thiamin structure: 4-amino-2-methylpyrimidine linked via CH₂–N⁺ to thiazole ring with CH₃ and CH₂CH₂OH substituents] — Thiamin

↓ SO_3^{-}

Pyrimidine sulphonic acid (4-amino-2-methyl-5-CH₂SO₃H pyrimidine) + Thiazole (with CH₃ and CH₂CH₂OH substituents)

After sulphite dipping and eight days of storage at 3 °C, whole and french-fried potatoes had lost 11% and 47%, respectively, of their original thiamin content (Anderson et al., 1954). Loss of thiamin in raw french fries analysed immediately after sulphite dipping was negligible, but increased to about 25% after several days of cold storage (Mapson & Wager, 1961). The extent of sulphite absorption depends upon the roughness of the surface of cut potatoes (Furlong, 1961) – the more damaged the surface, the greater the uptake. After dipping for only 5 min, most of the sulphite (33 p.p.m.) was concentrated in the outermost layer of the tuber, with only 0.3 p.p.m. in the rest of the tuber (Mudambi & Hanning, 1962). However, soaking for 16 h allowed sulphite penetration to a depth of 10 mm from the surface, as evidenced by a 55% destruction of thiamin at this level (Oguntona & Bender, 1976). This length of time of immersion of potato halves in either a metabisulphite solution or water showed that thiamin losses were approximately twice as great in the former as in the latter (Oguntona & Bender, 1976). Moreover, although this treatment was much more drastic than normal commercial processes, two out of five commercial samples of pre-peeled raw french fries analysed had low contents of thiamin (mean value 0.016 mg/100 g), which were similar to the laboratory-prepared samples (mean value 0.015 mg/100 g).

Previous treatment with sulphite increases losses caused by cooking (Mapson & Wager, 1961; Table 4.4). Other authors have found similar losses. Sulphiting doubled the loss caused by frying from 10% in half-tubers soaked in water to 20% in those soaked in sulphite (Oguntona &

Table 4.4. *Losses of thiamin due to sulphiting*[a]

Treatment	% Total losses of thiamin
Unpreserved	—
Preserved[b], analysed immediately	0
Unpreserved, whole boiled	17
Preserved[b], whole boiled	32
Unpreserved, french fried	10
Preserved[b], french fried	44
Preserved, french fried and kept hot for 18 min before serving	72
Preserved[b], stored 1–7 days at 5 °C	21–27

[a] Figures taken from Mapson & Wager (1961).
[b] Dipped in 1.0% sodium metabisulphite solution for 2 min and drained for 2 min.

Bender, 1976). Boiling after sulphiting produced a loss of about 30% thiamin in three varieties of halved tubers, compared to about 20% caused by boiling alone, according to Mudambi & Hanning (1962). A 100 g serving of boiled, untreated potatoes would therefore provide 0.08 mg of thiamin compared to 0.07 mg for boiled, sulphited, a difference these authors considered unimportant in terms of daily requirements. It should be noted, however, that they did not store the sulphited potatoes before cooking and the losses they found through sulphiting alone were low (only 4% to 10%).

In general, findings indicate that sulphiting, when followed by storage for several days and subsequent cooking, significantly reduces the thiamin content of potatoes.

Frozen potato products

Since 1970, frozen potato products have accounted for 45% to 48% of all potatoes used for processing in the USA, or nearly one-quarter of that country's food use of potatoes (Weaver *et al.*, 1975).

Frozen french fries (frozen chips)

These are the most important of the frozen potato products (Figure 4.9). They may be either par-fried or finish-fried by the processor; in the former case they are later finish-fried in deep fat, and in the latter, oven-heated before consumption. A description of the production of frozen french fries and other frozen products has been given by Weaver *et*

124 *Effects of storage, cooking and processing*

al. (1975) and Smith (1975). Figure 4.10 is a flow diagram showing unit operations involved in production. Nitrogenous constituents, vitamins and minerals are all affected by these operations.

Nitrogenous constituents

Calculations of crude protein contents given by Murphy *et al.* (1966) on a moisture- and fat-free basis for home-prepared fries and commercial brands of frozen french fries show that, on average, commercial brands contained only 66% of the protein of the home-prepared item. Investigating effects of the main production operations involved with exposure of potatoes to water or high temperature (Augustin *et al.*, 1979b) indicated that total N was reduced to 85% and 81%, respectively, of its original value in large- and small-sized french fries, the main point of loss being blanching in water. Significant ($P < 0.05$) losses of glutamic and aspartic acids, valine, phenylalanine, arginine, methionine and tryptophan occurred during hot-water (77 °C) blanching of 0.95 cm-thick french fries during an experimental process simulating commercial practice (Kozempel *et al.*, 1982). Most of the losses were suggested to result from leaching of free amino acids into the blanch water.

Additional losses of amino acids occur through Maillard browning during frying (Fitzpatrick *et al.*, 1965), although only to a small extent in french fries (Jaswal, 1973). Losses in protein-bound and free amino acids were 5% and 7%, respectively, in the case of low specific gravity (LSG),

Figure 4.9. Millions of pounds of frozen french fries are processed at the J. R. Simplot Company plant at Caldwell, Idaho. (Photo courtesy of J. R. Simplot Company.)

Processing

and 4% and 10% for high specific gravity (HSG), potatoes. Furthermore, availability of lysine decreased by 12% (LSG) and 14% (HSG) on frying. The author did not investigate the cause of these changes, but assumed they were due to amino acid–carbohydrate interactions.

Vitamins

Vitamin losses during the commercial production of french fries can be considerable. The processes leading to greatest vitamin losses are peeling and slicing (Gorun, 1978*b*) and blanching (Gorun, 1978*b*; Augustin *et al.*, 1979*b*). The frying operation was suggested to have little effect on vitamin content. Other workers, however, encountered losses in total ascorbic acid (Pelletier *et al.*, 1977) and thiamin (Oguntona & Bender, 1976), which are due to frying alone, albeit under laboratory conditions.

Other factors influencing extent of vitamin losses are: previous storage of the raw material, size of french fry cut, type of blanching (steam or water), and finishing operations (frying or oven-heating). No investigation into effects of freezing on nutrient content of frozen french fries has been reported.

Overall losses in the finished product, when freshly harvested potatoes were used, were 44% ascorbic acid, 44% thiamin, 39% riboflavin and 24% niacin (Gorun, 1978*b*). These increased to 72%, 52%, 45% and 35%, respectively, when the potatoes were stored for six months prior to use. Somewhat lower losses for large (0.5 in.) and small (0.25 in.) french

Figure 4.10. Processes for making frozen french fries and patties.

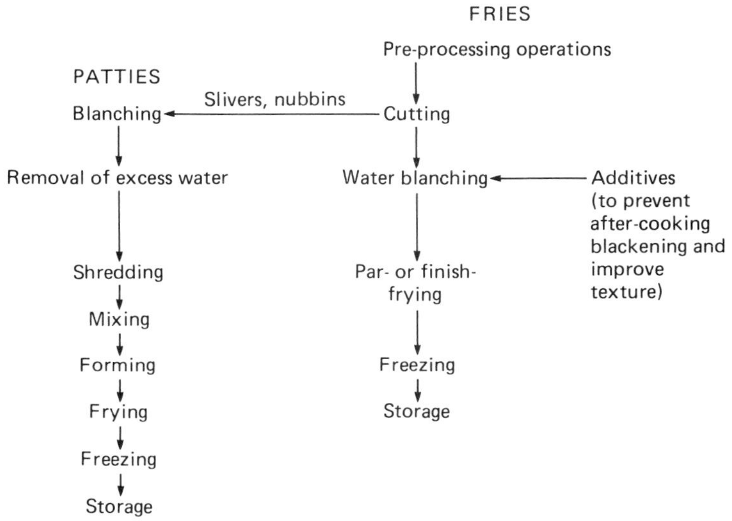

```
                                              FRIES
                                   Pre-processing operations
         PATTIES                              ↓
                    Slivers, nubbins
         Blanching ←─────────────── Cutting
             ↓                          ↓
  Removal of excess water         Water blanching ←──────── Additives
             ↓                          ↓                   (to prevent
             ↓                          ↓                   after-cooking
                                                            blackening and
          Shredding                Par- or finish-          improve
             ↓                        frying                texture)
           Mixing                       ↓
             ↓                          ↓
          Forming                    Freezing
             ↓                          ↓
           Frying                    Storage
             ↓
          Freezing
             ↓
          Storage
```

fries were reported by Augustin et al. (1979b) for potatoes stored for two to six months: reduced ascorbic acid, 31% and 39%; thiamin, 20% and 19%; niacin, 16% and 26%; pyridoxine, 22% and 26%; folic acid, 34% and 35%. It is clear that the most heat-sensitive vitamins – ascorbic and folic acids – are the least retained during processing.

A comparison of blanching methods (Augustin et al., 1979b) in small-size french fries showed significantly better retentions for reduced ascorbic acid, thiamin, niacin, pyridoxine and folic acid in the case of steam blanching (89% to 97%) than for water blanching (66% to 88%), presumably due to greater leaching losses in the latter process. The same study compared vitamin retentions in small- (0.6 cm) and large-sized (1.25 cm) french fries. With the exception of thiamin and folic acid, retention values were significantly lower in small than in larger fries (see figures quoted above). The authors attributed this to differences in leaching losses due to greater ratio of surface area:volume in the smaller-sized cuts. Blanching times and temperatures can also affect the extent of losses. Retentions of ascorbic acid in 1.25 cm fries were significantly reduced from 83% after blanching for 5 min at 66 °C to 54% after 15 min at 88 °C (Artz et al., 1983). Significant losses in ascorbic acid, thiamin, niacin and riboflavin, which increased with increasing hot-water (77 °C) blanch times of 4 to 20 min were recorded for 0.95 cm-thick french fries by Kozempel et al. (1982). Assuming losses to be due to leaching, the authors described a leaching model, with diffusion as the rate controlling step, and successfully predicted losses of these vitamins as a function of process parameters. There was a greater loss of ascorbic acid than of the B vitamins during blanching.

Pelletier et al. (1977) discovered that commercial samples of frozen french fries, purchased from Canadian food stores over two years and finish-cooked according to producers instructions, averaged 30% less total ascorbic acid than freshly prepared french fries. Total ascorbic acid content of french fries finished by cooking in oil averaged 2 mg/100 g higher than those finished by heating in an oven.

Significant amounts of all vitamins are lost during the preparation of frozen french fries. Processing losses are chiefly due to peeling, and operations, such as cutting and blanching, which lead to leaching. Some loss of vitamins may also result from freezing and reheating, but these have not been investigated.

Minerals

No information is available about losses of minerals during frozen french fry production. However, it seems safe to assume that losses will occur through removal of the outer tuber layers during peeling and as a result of leaching during slicing and blanching. Figures of Murphy *et al.* (1966) for ash contents of home prepared and seven commercial brands of fries can be used to calculate ash on a moisture- and fat-free basis. In this way it can be shown that commercial brands averaged only about 70% of the ash content of home-prepared fries. Thus 30% ash must have been lost during processing and finishing operations other than deep-fat frying.

Other frozen products

Potato patties, puffs and rounds, hash-browns and mashed potato are other frozen products. Information is scant on nutrient losses in these products, apart from one study (Augustin *et al.*, 1979*b*) on potato patties (for production operations see Figure 4.6). Retentions for total N (90%), reduced ascorbic acid (53%), thiamin (88%), niacin (90%), pyridoxine (91%) and folic acid (73%), respectively, were found in commercial patties. Unlike the case of french fries, these losses were not due mainly to blanching, a surprising result in view of the fact that, at the blanching stage, the patties are unformed and consist of slivers and pieces smaller than most french fries. The authors assumed that losses were due to leaching into the water to which the pieces of potato are extensively exposed prior to blanching.

In contrast to the high retention of total N found in this study, it can be calculated, using figures given by Murphy *et al.* (1966), that commercial brands of hash-browns, patties and puffs contained only 50%, 47%, and 40%, respectively, of the total N of the corresponding home-prepared items. Considerable quantities of N, therefore, must have been lost during commercial production.

The effect of freezing on contents of some nutrients has been studied in relation to production of frozen potato products in general (Mondy & Chandra, 1979*a,b*). Evidence indicates that freezing can cause a loss of ascorbic acid, although the extent of loss is probably dependent on the rate of freezing. During commercial processing, potatoes are frozen rapidly, preventing cell damage and consequent thawing loss of soluble compounds. Frozen–thawed mashed potatoes (prepared from instant potato flakes) lost only about 7% of their reduced ascorbic acid compared to the same item freshly prepared, and there were no losses of either riboflavin or thiamin (Ang *et al.*, 1975). A similar loss (6%) of total ascorbic acid resulted from rapid freezing and thawing of fresh mashed potatoes (Jadhav *et al.*, 1975).

Large losses of 43% and 22% reduced ascorbic acid in 'Katahdin' and 'Atlantic' cultivars were found (Mondy & Chandra, 1979*a*) as a result of slow freezing (−20 °C for 24 h), a factor which would have increased freezing damage and hence the exudation of juice, containing ascorbic acid, upon thawing. The same rate of freezing produced significant losses of potassium, phosphorus, calcium, magnesium, iron, copper, and zinc from both the cortex and pith regions (Mondy & Chandra, 1979*b*). The relevance of these two studies to commercial processing is somewhat dubious and they are probably more applicable to domestic freezing. Riemschneider *et al.* (1976) found a loss of more than 50% ascorbic acid in cooked potatoes which had been stored frozen for four months and quote another source which noted a 56% loss after six months of frozen storage. Products to be sold to the public are also kept stored for varying times in retail freezing cabinets, whose temperatures can fluctuate considerably causing changes in nutritional value. Strachan (1983) has indicated that quality (including nutritive value) of the frozen foods is subject to changes right through the distribution chain from storage of raw materials before processing up to consumption.

Reheating of frozen products can lead to yet further losses of vitamins. Ang *et al.* (1975) found reduced ascorbic acid retentions of only about 24% to 36% (depending upon the reheating method) in instant mashed potatoes frozen and reheated for half an hour. These retentions were similar to that of the freshly prepared item warm-held for 3 h, during which time its temperature rose from about 65 °C to 79 °C. Riboflavin was little affected, but thiamin was reduced to only about 88% of the value of freshly prepared instant mashed potatoes. The extent of total nutrient loss in frozen potato products has not been investigated systematically and much work remains to be done, especially since the popularity of such products has increased greatly in recent years.

Potato chips (crisps)

The potato chip is a fried snack product which until recently was the most important form of processed potato in the UK. In 1978/79 a total of 38% of all potatoes destined for UK processing went to the chips sector, although in the same years the french fry superseded the chip in terms of importance (Young, 1981). (For a detailed description of chip manufacture, see Smith, 1975.)

A flow diagram illustrating the operations involved in the production of chips is shown in Figure 4.11. The operations are similar to those for preparation of french fries; however, chip potatoes are cut into very thin slices, rather than rods. Frying greatly reduces moisture content (to about 2%) in the finished chips, which are stored in sealed bags, not frozen.

Nitrogenous compounds

Studies of the effect of chipping on the nitrogenous compounds of potatoes show that there is considerable damage to the protein. Fitzpatrick *et al.* (1965) and Fitzpatrick & Porter (1966) found losses of free amino acid N amounting to 50% to 60% when fresh potatoes were fried, with a corresponding loss of about 70% of the reducing sugars. Accumulation of reducing sugars during cold storage increased losses of free amino acids on frying to 85% to 88%. Analyses of individual amino acids revealed decreases ranging from slight to large in all of the amino acids reported, even when fresh potatoes were fried. Losses of all amino acids were great (100% of methionine) in stored fried chips; only part of the lost methionine was oxidized to methionine sulphoxide during frying. At least some losses were due to the reaction of amino acids and sugars during Maillard browning; however, when potatoes which had previously been cold-stored were reconditioned and their reducing sugar content greatly lowered, a large loss of amino N took place on frying. The reason for this loss was not known. Jaswal (1973) also studied the effect of chipping on the amino acids of LSG and HSG potatoes. His results can be summarized as follows:

	% Lys loss		% Amino acid loss	
	Total	Available	Protein-bound	Free
LSG	58	67	37	45
HSG	38	62	20	33

The losses of total and available lysine are particularly marked.

Figure 4.11. Processes for making chips (crisps).

Pre-processing operations
↓
Slicing and washing
↓
Removal of moisture
↓
Frying
↓
Salting and flavouring
↓
Packing

Ascorbic acid

The only vitamin studied in relation to effects of chipping is ascorbic acid. In the case of other vitamins it is likely that losses during chipping will at least equal those during french frying (see pp. 125–6) and are probably greater. Bucko et al. (1977) found that ascorbic acid losses during frying were considerably higher when potatoes were cut into thin slices than when they were in the form of rods. It is likely that all vitamins will be similarly affected, due to the increased surface area exposed to the effects of leaching and heat-destruction during processing. Pelletier et al. (1977) report losses of 30% to 85% ascorbic acid in the preparation of chips, as recorded by various authors. They themselves estimated average loss of total ascorbic acid in commercial Canadian brands to be about 75%.

Minerals

No information is available about mineral losses during chipping. However, as in the case of vitamins, it seems likely that losses through leaching may be substantial during washing and blanching, due to the greater surface area exposed to the water.

Other observations

In spite of substantial losses of nitrogenous compounds and vitamins during processing, chips are still quite a good source of these nutrients, as a considerable concentration effect is achieved through reduction of moisture content (Table 4.7). However, the nutritional quality of retained N has not been studied. Absorption of large quantities of fat together with an insignificant reduction in carbohydrate during processing ensure that chips are a highly concentrated form of energy. Weaver et al. (1983) consider that the estimate of 33.3 g of chips eaten as a single serving is more reasonable than the 100 g serving used for other potato products. Table 4.8 compares the content of some nutrients in 33.3 g of chips with those in 100 g of raw, cooked or processed potato with a similar potato solids content. In general, 33.3 g of chips contain a lower concentration of nutrients than did 100-g servings of boiled, peeled, baked or french-fried potatoes, but a greater concentration than did 100 g of rehydrated flakes or granules. The value of chips as a snack food has been described enthusiastically by Deutsch (1978).

One group of workers has suggested production of potato chips from unpeeled potatoes as a way of increasing the yield and nutritive value of the finished product, and decreasing waste disposal problems (Shaw et al., 1973). There was no significant difference in flavour or appearance of chips from peeled or unpeeled potatoes, as judged by a panel of trained tasters. However, the use of unpeeled potatoes to produce chips would have to be subject to the assurance that any glycoalkaloids present were not concentrated to toxic levels (see Chapter 5, p. 176).

Dehydrated potato products

Dehydration is one of the major means of preserving potatoes, giving products such as potato flour, granules, flakes and dice. The dehydration industry is particularly important in France, using 61% of all potatoes destined for processing in 1977/78 (Young, 1981).

Potato flour can be incorporated into bread and is used as a combination thickener–flavouring agent in products such as dehydrated soups, gravies, sauces and baby foods. Potato granules and flakes are convenience foods for both domestic and large-scale use. The granule and flake processes can use potatoes rejected by other sectors, as they are less sensitive to raw material requirements (Hughes, 1983). Dehydrated potato dice are ingredients in processed foods such as canned meats, meat stews, frozen meat pies and potato salad.

The processes leading to production of potato granules (Boyle, 1975), flakes (Willard & Kluge, 1975), dice (Kueneman, 1975) and flour (Willard, 1975) have been reviewed elsewhere and described in detail by Feustel *et al.* (1964). Figures 4.12 and 4.13 are flow diagrams of the processes.

Nitrogenous constituents

Changes in potato nitrogenous constituents during dehydration have been demonstrated both during a commercial process and in laboratory simulations of these operations. These changes, as in the case of fried products, are of two types: physical losses due to peeling and leaching, and chemical losses as a result of amino acid–sugar interactions at the elevated temperatures required for drying. During a small-scale process to produce dehydrated diced potato, peeling increased the ratio crude:pure protein, indicating that peeling removed more pure protein than NPN (Kempf *et al.*, 1976). The same ratio decreased considerably during hot-water blanching, showing that more NPN than pure protein had been lost through leaching into the water. Dehydration itself produced no further change in the ratio, as might be expected, since losses of free amino acids as a result of chemical reactions were not studied. During processing, water blanching greatly affected total N in granules, flakes, slices and dices (Augustin *et al.*, 1979*a*). Dehydration had little effect on nitrogen content in the case of granules, slices and dices, but considerably

132 Effects of storage, cooking and processing

reduced it in the case of flakes. The overall retention values for total N were 83% (granules), 70% (flakes), 85% (slices) and 86% (dices). Servings of mash made from flakes or granules had only 65% to 70% of the total N found in an equal weight of boiled potatoes (Weaver *et al.*, 1983).

Jaswal (1973) determined total losses of free amino acids, as a result of drum-drying in the laboratory to be about 19% in the case of LSG potatoes (a loss greater than that of french fries, but only about half that of chips). There was a slightly greater loss from the free amino acid than from the protein-bound fraction. Losses from HSG potatoes were small (only about 4%), presumably because these potatoes were in contact with the high temperature drum surface for a shorter time than were the LSG potatoes to reduce them to the same moisture content, although the reason for the difference was not stated. There was a loss of available lysine of about 21%, with a greater loss from the free, than from the bound, fraction. These losses were assumed to be the result of reactions between amino acids and carbohydrate.

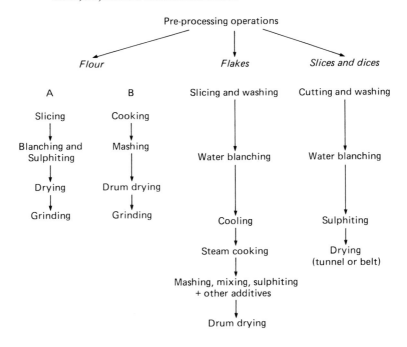

Figure 4.12. Processes for producing dehydrated products. A, a method used by Roy Choudhuri *et al.* (1963a), suitable for simple processing (e.g. rural). B, normal industrial method.

Processing

Maga & Sizer (1979), however, found high losses of free amino acids as a result of browning reactions during the commercial processing of potatoes into drum-dried potato flakes. The amino acid most affected was methionine, 85% of which was lost. When the potato flakes were re-wetted and extruded at high temperatures, the losses of all free amino acids were further increased.

Vitamins

Investigations into total or reduced ascorbic acid contents of unenriched commercial brands of dehydrated products have almost invariably shown low levels of the vitamin (Alarcón, 1977; Hanning & Mudambi, 1962; Myers & Roehm, 1963; Pelletier et al., 1977). Differences in reduced ascorbic acid between brands and between individual products within a brand have been reported (Myers & Roehm, 1963). Although these differences undoubtedly partially reflect differences in contents of raw materials, it has been found that processing greatly reduces ascorbic acid levels. The retentions of reduced ascorbic acid and

Figure 4.13. Processes for producing dehydrated granules.

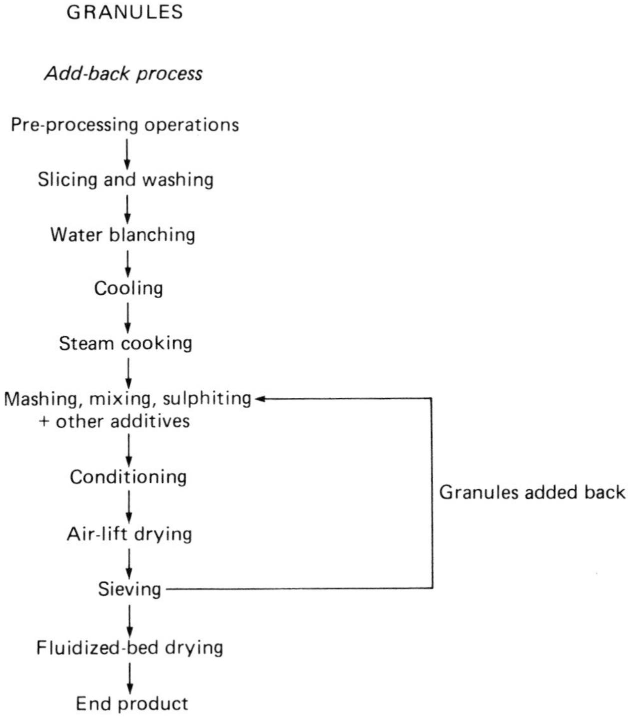

GRANULES

Add-back process

Pre-processing operations
↓
Slicing and washing
↓
Water blanching
↓
Cooling
↓
Steam cooking
↓
Mashing, mixing, sulphiting ←
+ other additives
↓
Conditioning | Granules added back
↓
Air-lift drying
↓
Sieving ─────────────
↓
Fluidized-bed drying
↓
End product

of other vitamins during the production of granules, flakes, slices and dices as found by Augustin et al. (1979a, 1982) are given in Table 4.5. Steele et al. (1976), however, showed total ascorbic acid losses in granules to be as high as 74%.

The manufacturer may compensate for these by enriching the final product with added vitamin C. However, in 1976 (two years after the amendment of the Canadian Food and Drug Regulations to permit ascorbic acid addition) an average of only 2.4 mg total ascorbate/100 g, was found in seven brands of commercial Canadian dehydrated potatoes (Pelletier et al., 1977), indicating that producers were not replacing lost ascorbic acid.

The critical operations affecting ascorbic acid loss during production of granules are water-blanching (Augustin et al., 1979a) and particularly mixing and mashing (Augustin et al., 1979a; Jadhav et al., 1975), presumably due to leaching and oxidation of the vitamin at the respective points. Further losses occur on conditioning, according to Jadhav et al. (1975). Dehydration itself apparently produces no further loss, in fact an apparent gain in total ascorbic acid was noted between the conditioning and final steps of processing (Jadhav et al., 1975). This was attributed to amino acid–sugar interaction products which interfere with ascorbic acid determinations in unmodified standard methods of analysis (Steele et al., 1976). Data based on these methods may hence be overestimating the final ascorbic acid content of dehydrated potato products. However, substantial losses of reduced ascorbic acid were recorded as a result of the dehydration process during the production of slices and dices (Augustin et al., 1979b) but not during that of flakes.

Table 4.5. *Final retentions of vitamins in commercial dehydrated potato products*[a]

Product	% Retention				
	Ascorbic acid	Thiamin	Niacin	Folic acid	Pyridoxine
Granules	45	9	78	48	83
Flakes	47	(63)[b]	77	54	62
Slices	40	4	73	58	72
Dices	38	4	80	69	84

[a] Figures from Augustin et al. (1979a).
[b] A later study by Augustin et al. (1982) found only traces of thiamin in potato flakes.

Thiamin values have also been severely reduced (see Table 4.5), in the case of granules, slices and dices to negligible proportions, during processing. Thiamin content in 100 g of each of five North American brands of reconstituted mashed potatoes was low and in one case negligible (Hanning & Mudambi, 1962). The major cause of thiamin destruction during processing is sulphite addition (Augustin *et al.*, 1979*a*). Flakes suffered least in this respect because they had been exposed for shorter times to both sulphite and high temperature drying conditions.

Some effects of processing on other vitamins are shown in Table 4.5. The vitamins most severely affected by commercial dehydration are thiamin (destroyed by sulphiting) followed by heat-sensitive ascorbic and folic acids. Retentions of niacin and pyridoxine are, however, well over 50%.

Reduction in folic acid has been shown to be a result of the combined effects of leaching during blanching, possible oxidation during mixing and mashing and, in the case of dehydrated slices and dices, destruction during dehydration (Augustin *et al.*, 1979*a*). Niacin, a heat-stable vitamin, was reduced largely by leaching during blanching, as was pyridoxine; the latter was also reduced by the dehydration step during flake production.

Additional losses of ascorbic acid have been noted during the storage of dried products. Roy Choudhuri *et al.* (1963*a*) found a loss of 25% to 30% ascorbic acid in three varieties processed as dried potato flour stored for six months at 37 °C in sealed tins. As much as 49% of the ascorbic acid which had been added to dried potato purée was lost during a storage period of three months at ambient temperature (Alarcón, 1977) in sealed plastic bags. It is possible that ascorbic acid acts as an antioxidant in dehydrated products, preventing lipid oxidation but being oxidized itself. Ascorbic acid losses should be controllable by the addition of antioxidants to the dried potato, or by packaging under vacuum to exclude oxygen.

There are also losses of ascorbic acid which take place during subsequent preparation and cooking of the dried products. Such losses were found by Myers & Roehm (1963) (who determined only the reduced form) to be high. Loss in dices prepared as hash-browned potatoes were about 37%; mashed potatoes prepared from flakes lost an average of 48% in one brand and 10% in another. About 70% of the ascorbic acid was lost when dried slices were prepared as fried potatoes. This diminished the final reduced ascorbic acid content in the cooked products to extremely low (almost negligible) values in some cases. Home preparation (simple reconstitution) of instant mashed potatoes from either flakes or granules, however, did not affect the contents of reduced ascorbic acid (Augustin *et al.*, 1982). When these preparations were subsequently warm-held for up to 60 min on a steam-table as is common practice in institutional feeding, the heat-sensitive vitamins, ascorbic and folic acids, were significantly reduced. The other vitamins were unaffected. Only traces of thiamin were detected even in the dehydrated products.

Chilling (to 5 °C) and reheating of mashed potatoes prepared from ascorbic acid-enriched granules caused small reductions in all the vitamins after 24 h of chilling and in all but pyridoxine after just 6 h of low temperature (Augustin *et al.*, 1980). Total retentions of vitamins in the dehydrated granules exceeded 90%, with the exception of ascorbic and folic acids (both at 86%). Thiamin levels were not reported.

Dehydrated potatoes are negligible sources of thiamin and may be poor sources of ascorbic acid unless enriched with the vitamin after processing. This could be a cause for concern if dried potatoes are substituted on a large scale for fresh potatoes, at certain times of the year, or at all times in certain sections of the population.

Minerals

There is only one published report (Roy Choudhuri *et al.*, 1963*a*) on loss of minerals during dried potato processing: during a simple process for the production of potato flour (see Fig. 4.8*a*), loss was attributed to leaching, presumably during the blanching and sulphite-dipping steps.

Canned potatoes

Canned food is convenient but it is bulkier and more expensive to transport than dehydrated products. In terms of total processed potato products, the canning sector plays a minor role. According to Talburt (1975*a*), in 1972 only about 3% of all potatoes destined to be processed in the USA were canned, either alone or with other ingredients. In some European countries the percentage is even lower (Young, 1981). Most canned potatoes are small and whole, but some may be diced, sliced or cut into strips. Their production has been detailed by Talburt (1975*b*).

Figure 4.14 shows operations involved in canning. As might be expected from the minor importance of canned potatoes, there is little information on changes in nutrient value during canning. The compositions of canned solids alone and of solids and liquids together are in Table 4.7.

The information available clearly shows that much nutrient loss from the potato solids themselves is actually due to a transfer into the surrounding liquid. At least one group of workers (Witkowski & Paradowski,

1975) has recommended the use of the canned liquid surrounding the potatoes as a supplement to soups and stews.

Protein

Total N of canned potatoes was less than that of the corresponding raw potatoes (Roy Choudhuri et al., 1963b), but the N lost (about 22%) was found to be present in the brine as a result of leaching during processing. However, significant losses of both bound and free amino acids (36% and 44%, respectively, in LSG and 19% and 25%, respectively, in HSG potatoes) were found on canning (Jaswal, 1973). It is possible that the author determined amino acids only in the potato solids, in which case these losses could have occurred at least partially by leaching into the brine around the potatoes. That heat damage also took place is shown by a reduction in the availability of lysine, about 40% in both LSG and HSG canned potatoes.

Vitamins

Ascorbic acid and thiamin are also partially lost from the potatoes themselves into the surrounding liquid. Hanning & Mudambi (1962) found approximately equal concentrations of either biologically available ascorbic acid or thiamin (mg/100 g) in the potatoes and in the liquid in several different brands of canned potatoes. In another study (Witkowski & Paradowski, 1975), up to one-third of the ascorbic acid passed from the potatoes into the surrounding brine. The loss from the potatoes increased with increasing durations and temperatures of heat treatment during

Figure 4.14. Processes involved in canning of potatoes.

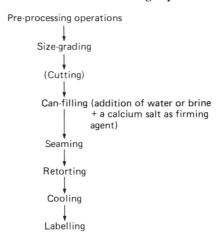

retorting. Reported losses vary considerably from about 6% to 8% (Witkowski & Paradowski, 1975) to 65% to 70% (Roy Choudhuri et al., 1963b).

Final concentrations of ascorbic acid in canned potatoes also ranged widely from rather low (5 to 6 mg/100 g; Roy Choudhuri et al., 1963b; Pelletier et al., 1977) to high (25 mg/100 g) in one brand tested by Hanning & Mudambi (1962). The latter workers also found wide ranges between different brands and even between different samples of one particular brand (e.g. 8.5 to 17.0 mg/100 g). The final concentration of ascorbic acid no doubt depends upon the initial concentration in the original raw material, the amount lost during processing operations other than heating, and the heating process itself.

Use of small, immature potatoes for canning could mean an initially high concentration in the raw material (Hanning & Mudambi, 1962). It is conceivable, therefore, that canned potatoes may be a better source of ascorbic acid than fresh potatoes that have been stored for long periods of time. However, if canned potatoes are subjected to cooking operations other than simple heating when removed from the can, they may undergo further losses of ascorbic acid. Canned potatoes stored chilled overnight and then browned, contained only 2 mg total ascorbic acid/100 g compared to 6 mg/100 g in those heated directly after removal from the can (Pelletier et al., 1977).

An average value of 0.037 mg thiamin/100 g drained potatoes was in various samples of eight brands of canned potatoes (Hanning & Mudambi, 1962), with quite wide variations between samples within brands. In general, however, values were low compared to values for boiled, fresh potatoes.

Figures for levels of other vitamins in drained canned contents given by Paul & Southgate (1978) show reductions, compared to the levels in raw potatoes, of approximately 25%, 50%, 30%, and 30% for riboflavin, niacin, pyridoxine and folic acid, respectively. It is impossible to say if these losses are due simply to leaching into the can liquid (as seems likely in the case of niacin) or are also caused partly by heat destruction.

Minerals

Ash content of canned potatoes was slightly higher than that of fresh potatoes (Roy Choudhuri et al., 1963b), probably due to absorption of sodium chloride from the surrounding brine. No difference was noted in iron or phosphorus contents between fresh and canned potatoes, and calcium content was only slightly higher in canned potatoes. This was probably due to the soaking of potatoes in calcium chloride, followed by

Processing

washing before blanching, filling and canning. It is possible that when – as is the practice in some canning plants – calcium salts are incorporated as firming agents into the liquid in the can, the calcium content of the final product may be significantly increased.

Comments on nutritional changes during processing

The approximate losses of vitamins occurring in processed products are given in Table 4.6 and can be compared with those occurring in home-cooked potatoes. In spite of such losses, it should not be assumed that all processed products are poor sources of nutrients. The nutrient content of equal weights of various processed products as determined by different authors are detailed in Table 4.7. These figures necessarily reflect differences in nutrient contents due to processing methods *per se*, those due to use of different varieties of potato, and those resulting from varying analytical methods of nutrient determination.

Table 4.6. *Approximate losses of vitamins during domestic preparation and processing*[a]

Method of preparation or processing	% Total loss of vitamins				
	Ascorbic acid	Thiamin	Niacin	Folic acid	Pyridoxine
Boiled, unpeeled[b]	20	10	0	20	0
Boiled, peeled[b]	20–50	0–40	0–30	10–40	15–20
Oven-baked[b]	25	15	5	30	10
Raw, fried[b]	30–50	10	5	20	—
Peeled, boiled, fried[b]	40	40	40	—	—
Mashed[b]	30–80	—	—	—	—
Hash-browned[b]	45–70	—	—	—	—
Salad[b]	65	—	—	—	—
Dumplings[b]	85	—	—	—	—
Pre-peeled, sulphited, boiled[c]	30	30	—	—	—
Pre-peeled, sulphited, fried[c]	—	45	—	—	—
French fries[c]	25–35	20–40	20	35	25
Chips[c]	30–85	—	—	—	—
Flakes[c]	50	>90	25	50	40
Granules[c]	55	>90	25	50	20
Canned[cd]	10–70	50	50	30	30

[a] Figures taken from various authors mentioned in the text.
[b] Domestic preparation.
[c] Processing.
[d] Losses to consumer are less if can liquid is also consumed.

140 *Effects of storage, cooking and processing*

Table 4.8 provides an example of a study which compared nutrient contents in equal servings of four processed products made from a single variety, 'Russet Burbank', taken from a single storage lot of tubers, all of which were analysed by the same methods. This table also provides a comparison between two types of domestic preparation and four types of processed products. Processes resulting in instant mashed potato in the form of granules or flakes in general have the lowest nutrient contents in comparison with an equal weight of raw or boiled potato. Their thiamin

Table 4.7. *Composition of processed forms of potatoes (per 100 g)*

Form of potato	Energy (kJ)	Energy (kcal)	Moisture (%)	Crude protein (g)	Fat (g)	Total carb. (g)	Dietary fibre (Crude fibre) (g)
Raw	335	80	78.0	2.1	0.1	18.5	1.5
Frozen french fries							
(fried)[a]	1217	291	73.1	3.0	18.9	29.0	3.2
(heated)[c]	920	220	52.9	3.6	8.4	33.7	(0.7)
Frozen mashed							
(heated)[e]	389	93	78.3	1.8	2.8	15.7	(0.4)
Chips[d]	2305	551	2.3	5.8	37.9	49.7	11.9[a]
Potato flour[a]	1469	351	7.6	8.0	0.8	79.9	(1.6)
Flakes (prepared: water, milk, fat added)[c]	389	93	79.3	1.9	3.2	14.5	(0.3)
Granules (prepared: water, milk, fat added)[c]	402	96	78.6	2.0	3.6	14.4	(0.2)
Canned (solids only)[a]	222	53	84.2	1.2	0.1	12.6	2.5
Canned (solids and liquid)[c]	184	44	88.5	1.1	0.2	9.8	(0.2)

[a] Paul & Southgate (1978).
[b] Pelletier *et al.* (1977).
[c] Watt & Merrill (1975).
[d] Mean of figures given by Paul & Southgate (1978) and Watt & Merrill (1975).
[e] Value varies according to ascorbic acid in raw material, processing method and length of storage of dehydrated product. May be much higher if products enriched.
[f] Paul & Southgate (1978); estimated figure.
[g] May contain added calcium salts as firming agents.
[h] Figures based on results of Hanning & Mudambi (1962).

Processing

contents are extremely low, as are their ascorbic acid contents, unless manufacturers enrich them with added vitamin C. In contrast, a product such as frozen french fries may be as good or even a better source of the B vitamins and minerals than boiled potatoes, if served in the same quantity as the latter. This occurs when the retained nutrient concentration effect caused by moisture loss is greater than the nutrient processing loss.

Practical effects of changes in some major nutrients during processing

Ash (g)	Ca (mg)	P (mg)	Fe (mg)	Thiamin (mg)	Riboflavin (mg)	Niacin (mg)	Folic acid (μg)	Pyridoxine (mg)	Ascorbic acid (mg)
1.0	9	50	0.8	0.10	0.04	1.5	14	0.25	20
—	11	77	1.0	0.09	0.02	2.1	11	0.39	12[b]
1.4	9	86	1.8	0.14	0.02	2.6	—	—	9[b]
1.4	25	42	0.6	0.06	0.04	0.7	—	—	4
3.1	39	135	2.0	0.20	0.07	4.7	20	0.89	17
3.7	33	178	17.2	0.42	0.14	3.4	—	—	19
1.1	31	47	0.3	0.04	0.04	0.9	—	—	5[e]
1.4	32	52	0.5	0.04	0.05	0.7	5[a]	(0.18)[a,f]	3[e]
—	11[g]	31	0.7	0.04[h]	0.03	0.7	11	0.16	13
0.4	4[g]	30	0.3	0.04[h]	0.02	0.6	—	—	19

Table 4.8. *Contents of some vitamins, minerals and trace elements in domestic preparations and processed products of 'Russet Burbank' tubers (per 100 g serving)*[a]

Potato preparation	Thiamin (mg)	Niacin (mg)	Ascorbic acid (mg)	Calcium (mg)	Iron (mg)	Copper (mg)	Zinc (mg)
Raw	0.08	1.44	10.3	9.4	0.37	0.11	0.19
Boiled (peeled)[b]	0.100 (125)[c]	1.09 (76)	10.5 (102)	9.8 (104)	0.38 (103)	0.09 (82)	0.26 (137)
Oven-baked[b]	—	1.44 (100)	—	7.8 (83)	0.53 (143)	0.08 (73)	0.29 (153)
French fried[d,e]	0.140 (175)	1.54 (107)	12.9 (125)	15.0 (160)	0.51 (138)	0.11 (100)	0.26 (137)
Chips[d,f]	0.030 (38)	0.86 (60)	6.41 (62)	6.7 (71)	0.33 (89)	0.04 (36)	0.16 (84)
Flakes[d]	0.018 (23)	0.57 (40)	2.94 (29)	5.4 (57)	0.30 (81)	0.04 (36)	0.15 (79)
Granules[d]	0.004 (5)	0.59 (41)	3.20 (31)	7.1 (76)	0.22 (59)	0.03 (27)	0.14 (74)

[a] Data from Weaver et al. (1983).
[b] Domestic-style preparation.
[c] Parentheses indicate percentage of the constituent in raw potato found in an equal weight serving of prepared potato.
[d] Processed.
[e] Frozen french fries finished in frying oil.
[f] 33.3 g serving; considered to be a more realistic estimate than 100 g for chips.

Table 4.9. *Percentages of adult recommended daily allowances provided by 100 g servings of processed potato products*[a]

Potato product	Crude protein	Thiamin	Niacin	Folic acid	Pyridoxine[b]	Ascorbic acid	Iron
Boiled in skin[c]	6	8	8	7	11	50	7–12
Frozen, mashed reheated	5	5	4	—	—	13	7–12
French fries, finish-fried	8	8	11	6	18	40	11–20
Chips[d]	5	6	8	3	13	19	8–14
Flakes (prepared)	5	0–3	5	—	—	17	3–6
Granules (prepared)	5	0–3	4	3	8	10	6–10
Canned (solids)	3	3	4	6	7	40	3–6

[a] Unless otherwise indicated, calculated from figures for processed potato products given in Table 4.7 as percentages of RDAs given by Passmore *et al.* (1974).
[b] As percentage of USRDA.
[c] Domestic preparation.
[d] A 33.3 g serving; considered to be a more realistic estimate of a single serving of chips.

are indicated in Table 4.9, which compares the percentages of RDA provided by 100 g of potato boiled in its skin and those provided by equal weights of some processed products. It can be seen that frozen french fries and chips provide similar quantities of protein, the B vitamins and iron as boiled potato. However, the nutritional quality of the protein of these products has not been studied and may be reduced, in relation to that of boiled potato, by the Maillard browning reaction, as may that of flakes and granules. Instant mashed potatoes from flakes and granules provide much smaller percentages of the RDA for the B vitamins and ascorbic acid than do boiled potatoes. The small percentages of RDA provided by canned potatoes are probably of little dietary importance, since the quantity of canned potatoes consumed is small compared to that of other products. Chips eaten as a snack food can supply significant percentages of the RDA of some B vitamins and iron.

Traditional processing

The inhabitants of high Andean areas of Peru and Bolivia process potatoes by centuries-old, traditional methods. One product, *chuño*, has been an article of commerce and domestic use in Peru since the earliest times, as evidenced by archaeological discoveries of *chuño* in pre-Columbian graves on the Peruvian coast. According to Salaman (1949), 'No stew was or is today thinkable without it [*chuño*], nor is a journey undertaken without carrying a supply of it.'

The method of production, a type of freeze-drying known as *chuñificación*, is made possible by the occurrence of severe frosts during the nights of June and July in the high Andean regions. These frosts alternate with high daytime levels of solar radiation and low levels of relative humidity. The temperatures at a site of *chuño* preparation in Bolivia varied from night-time minima of −2 °C to −14 °C to day-time maxima of 26 °C to 35 °C (Rodriguez, 1974). The morning relative humidity was 15% to 39% and fell to only 12% to 18% at midday.

The products of *chuñificación* can take two forms: *chuño blanco* (white *chuño*), which is also known as *moraya* or *tunta*, and *chuño negro* (black *chuño*). A further type of processing employed both in the high Andes and in Peruvian coastal regions results in a product known as *papa seca* (dried potato), produced by boiling, peeling and sun-drying potato tubers. These products are shown in Figure 4.15.

The *chuño* processing methods facilitate the consumption of bitter varieties of potato which, owing to their frost resistance, can be grown at high altitudes. These varieties contain high levels of glycoalkaloids (see Chapter 5) and are not eaten as fresh potatoes except in times of dire need, such as a famine. Christiansen (1977) found that the glycoalkaloid level was reduced from about 30 mg/100 g in fresh bitter potatoes to about 4 mg/100 g and 16 mg/100 g in *chuño blanco* and *chuño negro*, respectively. Damaged or diseased tubers of non-bitter varieties, unsuitable for fresh consumption, may also be processed by the same methods. *Papa seca* is made exclusively from non-bitter potato varieties.

A further reason for processing is to preserve potatoes for long periods of time. *Chuño* can be kept unchanged for several years. According to Werge (1979), this is particularly important in the high marginal areas where bitter potatoes are grown, the irregularity of yields due to frost, storms and drought creating a necessity for foods that can be stored from year to year. In such instances, up to 80% of the diet may consist of *chuño* stored from previous harvests (Christiansen, 1977). Moreover, *chuño* is light and easily transported and can be ground into a flour if necessary.

Production of *chuño*

Processing of *chuño* will be described briefly. More detailed descriptions of the preparation of *chuño negro* and *chuño blanco* are given by Guevara Velasco (1945) and of *chuño blanco* and *papa seca* by Werge (1979).

Processing **145**

Chuño blanco

On a night when a particularly heavy frost is expected, potatoes are spread evenly on the ground over a previously selected site. The following morning a careful examination of the tubers determines whether they have been frozen. They are exposed to additional nights of frost if the freezing is incomplete. The frozen tubers then thaw during the day as the temperature rises, but they may be covered with straw to

Figure 4.15. Finished dried samples of *chuño blanco*, *chuño negro* **and** *papa seca*.

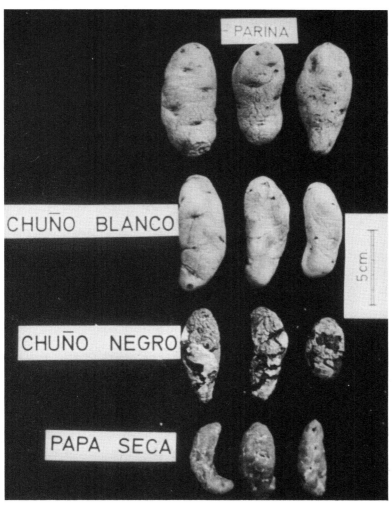

protect them from blackening by exposure to the sun's rays. Successive freezing and thawing periods cause tuber cells to separate, and destruction of the differential permeability of the cell membrane allows cell sap to diffuse from the cells into intercellular spaces (Treadway *et al.*, 1955).

Figure 4.16. Treading previously frozen and thawed potatoes for processing *chuño* in Peru.

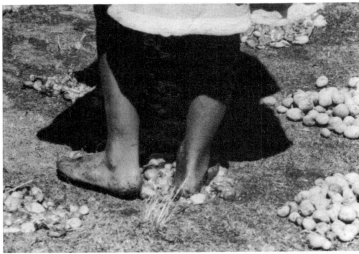

Processing   147

This released liquid is squeezed out of the tubers by trampling, a procedure which also removes the tuber skin (Figure 4.16). The trampled tubers are then transferred to a running stream (Figure 4.17) and immersed, covered with a protective layer of straw, for one to three

Figure 4.17. Transferring trampled potatoes to a stream for processing *chuño blanco* in Peru (top), and their subsequent removal after two to three weeks of soaking (bottom).

weeks. After removal from the water (Figure 4.17) they are spread in the sun to dry. The white crust which forms on the drying tubers gives the food its name.

A variation of *chuño blanco*, known as *tongosh* or *tokosh*, is prepared in some Andean areas. Tubers are soaked in water for up to a month, without prior freezing or pressing, and are in an advanced state of decay when removed from the water. They are subsequently dried and have a strong and distinctive odour when boiled.

Chuño normally forms the basis for stews and soups. When steam-heated with cheese, it is considered to be a particular delicacy. Mixed with fruit and molasses, it is eaten as a sweet dessert called *mazamorra* (Werge, 1979).

Chuño negro

Chuño negro production is similar to that of *chuño blanco*, except that tuber skins are not removed during trampling to squeeze out the juice and the trampled tubers are not soaked in water. After trampling they are immediately sun-dried, and the product is dark brown to black in colour.

Chuño negro is soaked in water for one to two days prior to cooking, to remove strong or bitter flavours, undesirable in the cooked item. Whole or as a flour, it is used chiefly in soups and stews.

Papa seca

Papa seca is produced by boiling and hand-peeling potatoes, which are then sliced or broken into small pieces and sun-dried. When dry, they are ground into finer pieces with a hand meat grinder. *Papa seca*

Table 4.10. *Composition of raw potato, chuño and papa seca per 100 g (FWB)*

Product	Energy (kJ)	Energy (kcal)	Moisture (%)	Crude protein (g)	Fat (g)	Carbo-hydrate (g)	Crude fibre (g)
Raw potatoes[a]	335	80	78.0	2.1	0.1	18.5	0.5
Chuño blanco[b]	1351	323	18.1	1.9	0.5	77.7	2.1
Chuño negro[b]	1393	333	14.1	4.0	0.2	79.4	1.9
Papa seca[b]	1347	322	14.8	8.2	0.7	72.6	1.8

[a] Sources listed in Table 2.2, p. 24.
[b] Collazos *et al.* (1974).

is used mainly for the preparation of a special dish called *carapulca* (a mixture of *papa seca*, meat, tomatoes, onions and garlic), but it may also be prepared as a soup. It is a higher prestige item than *chuño*, being eaten in the large cities and along the coast of Peru as well as in the mountains. *Chuño* consumption is largely confined to the high Andean areas.

Nutritive value of traditionally dried potato products

There is virtually no information on the nutritional value of *chuño* and *papa seca*, apart from that given in composition tables of Peruvian foods (Collazos *et al.*, 1974; Table 4.10). A better appreciation of the differences in nutrient concentration between raw and traditionally processed forms of potatoes is gained if nutrient contents are expressed on a dry weight basis, as in Table 4.11. This shows that, with the exception of carbohydrate, calcium and iron, nutrient contents of *chuño blanco* are greatly reduced in comparison with fresh potato. Those of *chuño negro*, including iron, are also reduced, but not to such a great extent as in *chuño blanco*. *Papa seca* has the highest nutrient content of the three forms, in terms of N and vitamins.

Christiansen (1977) demonstrated the major points of loss of total N during a laboratory simulation of the processes to produce *chuño blanco*, *chuño negro* and *papa seca* from three clones of *Solanum juzepczuckii* bitter potatoes. This information is summarized in Table 4.12.

Major points of nitrogen loss in *chuño blanco* are pressing (trampling) to extract juice, and soaking in water, the losses by the latter operation being much greater. In *chuño negro*, losses were due solely to pressing out of the exudate formed by freezing and thawing and there was only a small decrease in N in *papa seca* as a result of boiling.

Ash (g)	Ca (mg)	P (mg)	Fe (mg)	Thiamin (mg)	Riboflavin (mg)	Niacin (mg)	Ascorbic acid (mg)
1.0	9	50	0.8	0.10	0.04	1.50	20
1.8	92	54	3.3	0.03	0.04	0.38	1.1
2.3	44	203	0.9	0.13	0.17	3.40	1.7
3.5	47	200	4.5	0.19	0.09	5.00	3.2

Table 4.11. Composition of raw potato, chuño and papa seca per 100 g (DWB)[a]

Product	Energy (kJ)	Energy (kcal)	Crude protein (g)	Carbohydrate (g)	Ca (mg)	P (mg)	Fe (mg)	Thiamin (mg)	Riboflavin (mg)	Niacin (mg)	Ascorbic acid (mg)
Raw potato	1523	364	9.5	84.1	41	227	3.6	0.45	0.18	6.82	90.9
Chuño blanco	1649	394	2.3	94.8	112	66	4.0	0.04	0.05	0.46	1.3
Chuño negro	1623	388	4.7	92.4	51	236	1.0	0.15	0.20	3.96	2.0
Papa seca	1582	378	9.6	85.2	55	235	5.3	0.22	0.11	5.87	3.8

[a] Calculated from figures given in Table 4.10.

Table 4.12. *Changes in crude protein (N × 6.25) during the production of traditionally processed Andean potatoes*[a]

Product	% Crude protein (DWB)					
	Initial content	Loss in exudate	Loss on boiling	Pressed tuber content	Removed by soaking in water	Final content
Chuño blanco	12.5	2.9	—	9.3	6.5	2.8
Chuño negro	12.4	2.9	—	9.2	—	9.6
Papa seca	12.3	—	1.6	—	—	10.7

[a] Means of values for three clones of *S. juzepczukii*, calculated from figures given by Christiansen (1977).

Table 4.13. *Dry matter, retention of N and biological value of the retained N in traditionally processed potato products*[a]

Product	% Original retained			BV of retained N
	DM	Total N	AIS-N[b]	
Raw potato	100	100	100	48
Chuño blanco	88	20	38	56
Chuño negro	96	75	67	53
Papa seca	107	95	111	61

DM, dry matter; BV, biological value.
[a] Figures from Christiansen (1977).
[b] Alcohol-insoluble N. This is a measure of true protein as NPN compounds are extracted by alcohol.

Christiansen (1977; and see Table 4.13) calculated retentions of total N and alcohol-insoluble N (AIS-N) on the basis of losses of DM in the three processes. He also assessed the BV of the N in the products, expressing it as a percentage of that of casein, according to the microbiological method of Ford (1960).

Preparation of *chuño blanco* resulted in greater losses of DM and AIS-N than in the other two processes. *Chuño negro* also lost AIS-N but retained about twice as much as did *chuño blanco*. The apparent gain in DM and AIS-N in *papa seca* reflect sampling error due to large variations in the DM of individual tubers. However, the loss of AIS-N in *papa seca* must have been small. Processing apparently resulted in a slight improvement in the BV of the remaining N in all three products. The author does not comment on the reason for this. In the case of the *chuños* this could have been due to proportionately greater losses of NPN than of protein N.

Information given by Christiansen (1977) for losses of N in the three forms of traditionally processed potato also probably accounts for losses in other soluble constituents, especially the vitamins. However, these have not been investigated. The generally low nutrient level in *chuño blanco* undoubtedly results from a combination of losses due to squeezing out of tuber juice during thawing and trampling, and leaching losses during submersion in running water. The relatively high levels of calcium and iron in *chuño blanco* (Table 4.11) have not been explained but may result from their absorption by tubers from the soaking water. *Chuño negro* retains more nutrients than *chuño blanco* because it is not soaked in water before drying. *Papa seca* undergoes little nutritional change, as potatoes are boiled intact in their skins before sun-drying. All three forms have extremely low concentrations of ascorbic acid, presumably as a result of loss of exudate as well as losses due to heat destruction and oxidation during sun-drying (*chuño*), and to boiling followed by sun-drying (*papa seca*). The levels of pyridoxine and folic acid have not been investigated.

Part 4: Summary

Storage, cooking and processing have, in general, deleterious consequences on the content of potato nutrients. However, effects differ in extent, depending on the conditions of storage, the method of cooking, the type of processing employed, and the particular nutrient.

The highest levels of nutrients are found in freshly harvested potatoes which have been cooked unpeeled and intact. Long periods of storage

Summary

deplete tubers of ascorbic and folic acids. When storage is followed by cooking that requires more than one stage of preparation (for example boiling and frying) or by processing, the potato as it is eventually eaten may contain low levels of ascorbic acid. Research into nutrient changes resulting from on-farm storage in developing countries has been neglected. Changes in nitrogenous constituents during controlled storage are small and rather random in nature. They have been studied mainly in the NPN fraction and their nutritional significance is unknown. Domestic cooking has little effect on potato N apart from a slight reduction in concentration as a result of leaching into the cooking water. Processing operations, such as frying, drum-drying and canning, which involve higher temperatures than those used in domestic cookery, can reduce amino acid levels considerably as a result of carbohydrate–amino acid reactions during Maillard browning. Such reductions are more marked in potatoes containing high levels of reducing sugars resulting from cold storage. Their effect on the nutritional value of potato N in products such as chips and potato flakes has not been ascertained and deserves investigation.

Cooking and processing enhance digestibility of potato starch, which is indigestible in the raw state. The extent of cooked potato starch digestibility in young children is, however, unclear and should be investigated in relation to the use of potato in infant weaning foods.

Crude fibre is little changed during storage unless tubers are immature, in which case it increases. Dietary fibre changes have not been investigated. Changes during cooking have scarcely been investigated quantitatively or qualitatively. Removal of peel reduces the dietary fibre content of the tuber (see Table 2.4, p. 28).

The vitamins adversely affected by controlled storage are ascorbic and folic acids. There is little change in the other vitamins, with the exception of pyridoxine, which increases significantly in concentration. Ascorbic and folic acids suffer large losses during cooking and processing as a result of leaching, heat destruction and oxidation. Thiamin is destroyed by sulphiting and is reduced to low levels in pre-peeled potatoes and to negligible proportions in some dehydrated products. Thiamin, riboflavin and pyridoxine are all subject to losses as a result of heat destruction, and leaching into cooking or blanching water, but these losses are less than those of ascorbic and folic acids. Niacin is heat stable and is lost during cooking and processing mainly by leaching into cooking or blanching water.

Reductions in mineral contents have been studied to a limited extent. Minerals concentrated in the outer parts of the potato tuber are lost on peeling, but the extent has been little studied. Tuber mineral contents do not change during storage. Losses during cooking and processing are presumed to be mainly due to leaching into the cooking or blanching water.

Figures provided in the tables relating to the contents of various cooked and processed forms of potato are to be regarded as averages only. The wide range of nutrient contents in the raw material results in a correspondingly wide range of contents in potatoes as eaten. It is perfectly possible therefore, for a processed product prepared from freshly harvested material of high nutrient content to be a better source of a particular nutrient than are domestically cooked potatoes which were previously stored for a long time. Final nutrient contents in cooked or processed potatoes therefore depend on a variety of factors including initial content in the raw starting material, manipulations involved in cooking or manufacturing process and consumer or institutional handling of the cooked or processed products in terms of re-heating and periods of holding after preparation.

Traditional products resulting from processing of bitter potatoes in the Andean highlands of Peru and Bolivia are reduced in nutrients, especially nitrogenous constituents and ascorbic acid, in relation to the raw starting material. Further research is required into nutrient changes taking place during processing, especially of *chuño*. Improvements in nutrient retentions whilst maintaining the desired reductions in glycoalkaloid concentration during *chuño* production could be of great benefit to marginal communities who depend upon supplies of *chuño* in times of fresh potato scarcity.

References

Alarcón, N. R. (1977). [Retention of ascorbic acid during dehydration of potatoes and enrichment of a commercial purè.] In Spanish. Thesis, Catholic University of Valparaiso, Chile.

Anderson, E. E., Esselen, W. B. & Fellers, C. R. (1954). Factors affecting the quality of pre-peeled potatoes. *Food Technol.* **8**: 569–73.

Ang, C. Y. W., Chang, C. M., Frey, A. E. & Livingston, G. E. (1975). Effects of heating methods on vitamin retention in six fresh or frozen prepared food products. *J. Food Sci.* **40**: 997–1003.

Anon. (1875). In *The complete works of Count Rumford*, vol. 5. Macmillan & Co., London.

Appleman, C. O. & Miller, E. V. (1926). A chemical and physiological study of maturity in potatoes. *J. Agric. Res.* **33**: 569–577.

Artz, W. E., Pettibone, C. A., Augustin, J. & Swanson, B. G. (1983). Vitamin C retention of potato fries blanched in water. *J. Food Sci.* **48**: 272–3.

Augustin, J., Johnson, S. R., Teitzel, C., Toma, R. B., Shaw, R. L., True, R. H., Hogan, J. M. & Deutsch, R. M. (1978*a*). Vitamin composition of freshly harvested and stored potatoes. *J. Food Sci.* **43**: 1566–70, 1574.

References

Augustin, J., Johnson, S. R., Teitzel, C., True, R. H., Hogan, J. M., Toma, R. B., Shaw, R. L. & Deutsch, R. M. (1978b). Changes in the nutrient composition of potatoes during home preparation. II. Vitamins. *Am. Potato J.* **55**: 653–62.

Augustin, J., Swanson, B. G., Pometto, S. F., Teitzel, C., Artz, W. E. & Huang, C.-P. (1979a). Changes in nutrient composition of dehydrated potato products during commercial processing. *J. Food Sci.* **44**: 216–19.

Augustin, J., Swanson, B. G., Teitzel, C., Johnson, S. R., Pometto, S. F., Artz, W. E., Huang, C.-P. & Shoemaker, C. (1979b). Changes in the nutrient composition during commercial processing of frozen potato products. *J. Food Sci.* **44**: 807–9.

Augustin, J., Toma, R. B., True, R. H., Shaw, R. L., Teitzel, C., Johnson, S. R. & Orr, P. (1979c). Composition of raw and cooked potato peel and flesh: proximate and vitamin composition. *J. Food Sci.* **44**: 805–6.

Augustin, J., Marousek, G. I., Tholen, L. A. & Bertelli, B. (1980). Vitamin retention in cooked, chilled and reheated potatoes. *J. Food Sci.* **45**: 814–16.

Augustin, J., Marousek, G. A., Artz, W. E. & Swanson, B. G. (1982). Vitamin retention during preparation and holding of mashed potatoes made from commercially dehydrated flakes and granules. *J. Food Sci.* **47**: 274–6.

Bantan, S., Krapež, M. & Vardjan, M. (1977). [Variation in ascorbic acid during development and storage of the tubers of potato cv. Vesna and Bintje.] In Slovene. *Biol. Vestn.* **25**: 1–4.

Booth, R. H. & Shaw, R. L. (1981). *Principles of potato storage*. International Potato Center, Lima.

Boyle, F. P. (1975). Dehydrated mashed potatoes – potato granules. In W. F. Talburt & O. Smith (eds.) *Potato processing*, 3rd edn. AVI Publishing Company, Inc., Westport, CT.

Bretzloff, C. W. (1971). Calcium and magnesium distribution in potato tubers. *Am. Potato J.* **48**: 97–104.

Bucko, A., Obonova, K. & Ambrova, P. (1977). [Effects of storage and culinary processing on vitamin C losses in vegetables and potatoes.] In German. *Nahrung* **21**: 107–12.

Burton, W. G. (1966). *The potato*, 2nd edn. Drukkerij Veenman BV, Wageningen.

Burton, W. G. (1974). Requirements of the users of ware potatoes. *Potato Res.* **17**: 374–409.

Burton, W. G. (1978). Post-harvest behaviour and storage of potatoes. In Coaker, T. H. (ed.), *Applied biology*, vol. 2. Academic Press, New York.

Chick, H. & Slack, E. B. (1949). Distribution and nutritive value of the nitrogenous substances in the potato. *Biochem. J.* **45**: 211–21.

Christiansen, J. A. (1977). The utilization of bitter potatoes to improve food production in the high altitude of the tropics. Ph.D. thesis, University of Cornell, Ithaca, NY.

Collazos, C. et al. (1974). [The composition of Peruvian foods.], 4th edn. [In Spanish.] Ministry of Health, Lima.

Davies, A. M. C. & Laird, W. M. (1976). Changes in some nitrogenous constituents of potato tubers during aerobic autolysis. *J. Sci. Food Agric.* **27**: 377–82.

Desborough, S. L. & Weiser, C. J. (1974). Improving potato protein. I. Evaluation of selection techniques. *Am. Potato J.* **51**: 185–96.

Deutsch, R. M. (1978). Science looks at potato chips. *Chipper Snacker* **35**: 15–19.

De Vizia, B., Ciccimarra, F., DeCicco, N. & Auricchio, S. (1975). Digestibility of starches in infants and children. *J. Pediatrics* **86**: 50–5.

Dreher, M. L., Dreher, C. J. & Berry, J. W. (1984). Starch digestibility of foods: a nutritional perspective. *CRC Crit. Rev. Food Sci. Nutr.* **20**: 47–51.

Effmert, B., Meinl, G. & Vogel, J. (1961). [Respiration, sugar level and ascorbic acid content of potato varieties at various storage temperatures.] In German. *Züchter* **31**: 23–32.

El-Harith, E. A., Dickerson, J. W. T. & Walker, R. (1976). Potato starch and caecal hypertrophy in the rat. *Food and Cosmetics Toxicol.* **14**: 115–21.

Englyst, H., Wiggins, H. S. & Cummings, J. H. (1982). Determination of the non-starch polysaccharides in plant foods by gas–liquid chromatography of constituent sugars as alditol acetates. *Analyst* **107**: 307–18.

Eppendorfer, W. H., Eggum, B. O. & Bille, S. W. (1979). Nutritive value of potato crude protein as influenced by manuring and amino acid composition. *J. Sci. Food Agric.* **30**: 361–8.

Faulks, R. M., Griffiths, N. M., White, M. A. & Tomlins, K. I. (1982). Influence of site, variety and storage on nutritional composition and cooked texture of potatoes. *J. Sci. Food Agric.* **33**: 589.

Feinberg, B., Olson, R. L. & Mullins, W. R. (1975). Pre-peeled potatoes. In W. F. Talburt & O. Smith (eds.), *Potato processing*, 3rd edn. AVI Publishing Company, Inc., Westport, CT.

Fernandez, J. & Aguirre, A. (1975). [Comparison of the amino acid content of potato tubers, preserved by different methods, after five months of storage.] In Spanish. *Anal. Bromatologia* **27**: 115–34.

Feustel, I. C. (1975). Miscellaneous products from potatoes. In W. F. Talburt & O. Smith (eds.), *Potato processing*, 3rd edn. AVI Publishing Company, Inc., Westport, CT.

Feustel, I. C., Hendel, C. E. & Juilly, M. E. (1964). Potatoes. In W. B. Van Arsdel and M. J. Compley (eds.), *Food dehydration*, vol. 2 *Products and technology*. AVI Publishing Company, Inc., Westport, CT.

Finglas, P. M. & Faulks, R. M. (1984). Nutritional composition of UK retail potatoes both raw and cooked. *J. Sci. Food Agric.* **35**: 1347–56.

Fitzpatrick, T. J. & Porter, W. L. (1966). Changes in the sugars and amino acids in chips made from fresh, stored and reconditioned potatoes. *Am. Potato J.* **43**: 238–48.

Fitzpatrick, T. J., Talley, E. A. & Porter, W. L. (1965). Preliminary studies on the fate of sugars and amino acids in chips made from fresh and stored potatoes. *J. Agric. Food Chem.* **13**: 10–12.

Ford, J. E. (1960). A microbiological method for assessing the nutritional value of proteins. *Br. J. Nutr.* **14**: 485–97.

Furlong, C. R. (1961). Preservation of peeled potato. 2. Uptake of sulphite by peeled and chipped potato treated with sodium metabisulphite. *J. Sci. Food Agric.* **12**: 49–54.

Goddard, M. S. & Matthews, R. H. (1979). Current knowledge of nutritive values of vegetables. *Food Technol.* **33**: 71–3.

Gorun, E. G. (1978a). [Effect of mode of potato peeling on contents of B-group vitamins.] In Russian. *Izv. Vyssh. Uchebn. Zaved., Pishch. Tekhnol.* no. 6: 154–5.

Gorun, E. G. (1978b). [Changes in the vitamin activity of potatoes during production of quick-frozen French fries.] In Russian. *Kholod. Tekh.*, no. 10: 15–17.

Guevara Velasco, A. (1945). [Local products from the industrialization of potato. *Chuño* and *moraya*.] In Spanish. *La vida agricola* **22**: 1011–24.

Habib, A. T. & Brown, H. D. (1957). Role of reducing sugars and amino acids in the browning of potato chips. *Food Technol.* **11**: 85–9.

Hadziyev, D. & Steele, L. (1976). Vitamin C increase in aerated potato slices. *Qual. Plant. Plant Foods Hum. Nutr.* **26**: 365–88.

Hampson, C. P. (1972). Potato processing trends, problems and research. *J. Roy. Agric. Soc. England* **133**: 19–26.

Hanning, F. & Mudambi, S. R. (1962). Dehydrated and canned potatoes. *J. Am. Diet. Assoc.* **40**: 211–13.

Harris, R. S. (1975). General discussion on the stability of nutrients. In R. S. Harris & E. Karmas (eds.), *Nutritional evaluation of food processing*, 2nd edn. AVI Publishing Company Inc., Westport, CT.

Hellendoorn, E. W., Van den Top, M. & Van der Weide, J. E. M. (1970). Digestibility 'in vitro' of dry mashed potato products. *J. Sci. Food Agric.* **21**: 71–5.

Hentschel, H. (1969). [The biological value of the potato in relation to different domestic methods of preparing potato dishes. 4. Vitamins and minerals.] In German. *Qual. Plant. Mater. Veg.* **17**: 201–16.

Herrera, H. (1979). Potato protein: nutritional evaluation and utilization. Ph.D. thesis, Michigan State University, MI.

Hughes, B. P. (1958). The amino acid composition of potato protein and of cooked potato. *Br. J. Nutr.* **12**: 188–95.

Hughes, J. C. (1983). Potato production and processing. *Chem. Ind.*, no. 15: 598–603.

Jadhav, S., Steele, L. & Hadziyev, D. (1975). Vitamin C losses during production of dehydrated mashed potatoes. *Lebensm.-Wiss. Technol.* **8**: 225–30.

Jaswal, A. S. (1973). Effects of various processing methods on free and bound amino acid content of potatoes. *Am. Potato J.* **50**: 86–95.

Johnston, F. B., Hoffman, I. & Petrosovits, A. (1968). Distribution of mineral constituents and dry matter in the potato tuber. *Am. Potato J.* **45**: 287–292.

Johnston, D. E. & Oliver, W. T. (1982). The influence of cooking technique on dietary fibre of boiled potato. *J. Food Technol.* **17**: 99–107.

Jones, G. P., Briggs, D. R., Wahlquist, M. L. & Flentje, L. M. (1985). Dietary fibre content of Australian foods. I. Potatoes. *Food Technol. Australia* **37**: 81–3.

Kempf, W., Fehn, K.-H. & Bergthaller, W. (1976). [Variations in the crude and pure protein contents during production of dehydrated potatoes for consumption.] In German. *Potato Res.* **19**: 357–70.

Klein, L. B., Chandra, S. & Mondy, N. I. (1982). Sprouting of potatoes: compositional changes in light and dark. *J. Food Biochem.* **6**: 97–109.

Kozempel, M. F., Sullivan, J. F., Della Monica, E. S., Egoville, M. J., Talley, E. A., Jones, W. J. & Craig, J. C. (1982). Application of leaching model to describe potato nutrient losses in hot water blanching. *J. Food Sci.* **47**: 1519–23.

Kubisk, A., Tomkowiak, J. & Andrzejewska, M. (1978). [The content of some trace elements in different parts of the tuber in five potato varieties.] In Polish. *Hodowla Rosl., Aklim. Nasienn.* **22**: 81–8.

Kueneman, R. W. (1975). Dehydrated diced potatoes. In W. F. Talburt & O. Smith (eds.), *Potato processing*, 3rd edn. AVI Publishing Company, Westport, CT.

Leichsenring, J. M. *et al.* (1951). Factors influencing the nutritive value of potatoes. *Minnesota Technical Bull.* no. **196**. University of Minnesota Agricultural Experiment Station, Minnesota.

Leichsenring, J. M., Norris, L. M. & Pilcher, H. L. (1957). Ascorbic acid contents of potatoes. I. Effect of storage and of boiling on the ascorbic, dehydroascorbic, and diketogulonic acid contents of potatoes. *Food Res.* **22**: 37–43.

Linnemann, A. R., van Es, A. & Hartmans, K. J. (1985). Changes in the content of L-ascorbic acid, glucose, fructose, sucrose and total glycoalkaloids in potatoes (cv. Bintje) stored at 7, 16 and 28 °C. *Potato Res.* **28**: 271–8.

López de Romaña, G., Graham, G. G., Mellits, E. D. & MacLean, W. C. (1980). Utilization of the protein and energy of the white potato by human infants. *J. Nutr.* **110**: 1849–57.

McCay, C. M., McCay, J. B. & Smith, O. (1975). The nutritive value of potatoes. In W. F. Talburt and O. Smith (eds.), *Potato processing*, 3rd edn. AVI Publishing Company, Westport, CT.

Maga, J. A. & Sizer, C. E. (1979). The fate of free amino acids during the extrusion of potato flakes. *Lebensm.-Wiss. Technol.* **12**: 13–14.

Mapson, L. W. & Wager, H. G. (1961). Preservation of peeled potato. 1. Use of sulphite and its effect on the thiamine content. *J. Sci. Food Agric.* **12**: 43–9.

Mareschi, J. P., Belliot, J. P., Fourlon, C. & Gey, K. F. (1983). Decrease in vitamin C content in Bintje potatoes during storage and conventional cooking procedures. *Int. J. Vitamin Nutr. Res.* **53**: 402–11.

Mazza, G., Hung, G. & Dench, M. J. (1983). Processing/nutritional quality changes in potato tubers during growth and long term storage. *Can. Inst. Food Sci. Technol. J.* **16**: 39–44.

Meiklejohn, J. (1943). The vitamin B_1 content of potatoes. *Biochem. J.* **37**: 349–54.

Míča, B. (1978a). [Changes in the contents of total nitrogen and protein nitrogen during the storage and boiling of potatoes.] In Czech. *Rostl. Vyroba* **24**: 861–7.

Míča, B. (1978b). [Effect of storage and boiling on the free amino acid content in potatoes.] In Czech. *Rostl. Vyroba* **24**: 731–7.

Míča, B. (1978c). [Changes in the contents of glucose, fructose, sucrose and lysine in potatoes during storage and boiling.] In Czech. *Rostl. Vyroba* **24**: 35–43.

Míča, B. (1979). [Contents of phosphorus and potassium during storage and boiling of potatoes.] In Czech. *Rostl. Vyroba* **25**: 71–6.

Mondy, N. I. & Chandra, S. (1979a). Quality of potato tubers as affected by freezing. I. Texture, ascorbic acid and enzymatic discoloration. *Am. Potato J.* **56**: 119–24.

Mondy, N. I. & Chandra, S. (1979b). Quality of potato tubers as affected by freezing. II. Lipids and minerals. *Am. Potato J.* **56**: 125–32.

Mondy, N. I. & Ponnampalam, R. (1983). Effect of baking and frying on nutritive value of potatoes: minerals. *J. Food Sci.* **48**: 1475–8.

Mondy, N. I. & Rieley, P. B. (1964). Relationship of specific gravity to the nitrogen and ascorbic acid content of potatoes. *Am. Potato J.* **41**: 417–22.

Mudambi, S. R. & Hanning, F. (1962). Effect of sulphiting on potatoes. *J. Am. Diet. Assoc.* **40**: 214–17.

Murphy, E. (1946). Storage conditions which affect the vitamin C content of Maine-grown potatoes. *Am. Potato J.* **23**: 197–218.

Murphy, E. W., Marsh, A. C., White, K. E. & Hagan, S. N. (1966). Proximate composition of ready-to-serve potato products. *J. Am. Diet. Assoc.* **49**: 122–7.

Myers, P. W. & Roehm, G. H. (1963). Ascorbic acid in dehydrated potatoes. *J. Am. Diet. Assoc.* **42**: 325–7.

Nankar, J. T. & Nankar, V. J. (1979). Studies of the processing and utilization of potato in rural India. In H. Kishore (ed.), *Post harvest technology and utilization of potato*. International Potato Center, Region VI, New Delhi.

Oguntuna, T. E. & Bender, A. E. (1976). Loss of thiamin from potatoes. *J. Food Technol.* **11**: 347–52.

References

Page, E. & Hanning, F. M. (1963). Vitamin B_6 and niacin in potatoes. *J. Am. Diet. Assoc.* **42**: 42–5.

Passmore, R., Nicol, B. M., Narayana Rao, M., Beaton, G. H. & De Maeyer, E. M. (1974). *Handbook on human nutritional requirements*, WHO Monograph Ser. no. **61**. FAO/WHO, Geneva.

Paul, A. A. & Southgate, D. A. T. (1978). *McCance and Widdowson's The composition of foods*, 4th edn, MRC special report no. 297. HMSO, London.

Pelletier, O., Nantel, C., Leduc, R., Tremblay, L. & Brassard, R. (1977). Vitamin C in potatoes prepared in various ways. *J. Inst. Can. Sci. Technol. Aliment.* **10**: 138–42.

Ponnampalam, R. & Mondy, N. I. (1983). Effect of baking and frying on nutritive value of potatoes: nitrogenous constituents. *J. Food Sci.* **48**: 1613–16.

Porter, W. L. & Heinze, P. H. (1965). Changes in composition of potatoes in storage. *Potato Handbk* **10**: 5–13.

Priestley, R. J. (1979). Vitamins. In R. J. Priestley (ed.), *Effects of heating on foodstuffs*. Applied Sci. Publishers Ltd, London.

Rastovski, A., Van Es, A. *et al.* (1981). *Storage of potatoes. Post-harvest behaviour, store design, storage practice, handling*. Centre for Agricultural Publishing and Documentation, Wageningen.

Riemschneider, R., Abedin, M. Z. & Mocellin, R. P. (1976). [Quality and stability evaluation of food preserved by heat utilizing vitamin C as a criterion.] In German. *Alimenta* **15**: 171–4.

Rodriguez, E. M. (1974). [Traditional preparation of chuñu in Bolivia.] In Spanish. *Revista de la Sociedad Boliviana de historia natural* **1**: 30–5.

Roine, P., Wichmann, K. & Vihavainen, Z. (1955). [The content and stability of ascorbic acid in different potato varieties in Finland.] In Finnish. *Suom. Maataloust. Seur. Julk.* **83**: 71–87.

Roy Choudhuri, R. N., Joseph, A. A., Joseph, K., Narayana Rao, M., Swaminathan, M., Sreenivasan, A. & Subrahmanyan, V. (1963a). Preparation and chemical composition of potato flour from some varieties of potato. *Food Sci. (Mysore)* **12**: 251–3.

Roy Choudhuri, R. N., Joseph, A. A., Ambrose Daniel, V., Narayana Rao, M., Swaminathan, M., Sreenivasan, A. & Subrahmanyan, V. (1963b). Effect of cooking, frying, baking and canning on the nutritive value of potato. *Food Sci. (Mysore)* **12**: 253–5.

Salaman, R. N. (1949). *The history and social influence of the potato*. Cambridge University Press, Cambridge. (Reprinted 1970.)

Schwerdtfeger, E. (1969). [The biological value of the potato in relation to different domestic methods of preparing potato dishes. 3. Protein and amino acids.] In German. *Qual. Plant. Mater. Veg.* **17**: 191–200.

Seiler, H., Schlettwein-Gsell, D., Brubacher, G. & Ritzel, G. (1977). [Mineral content of potatoes in relation to the method of preparation.] In German. *Mitt. Geb. Lebensmittelunters. Hyg.* **68**: 213–24.

Shaw, R., Evans, C. D., Munson, S., List, G. R. & Warner, K. (1973). Potato chips from unpeeled potatoes. *Am. Potato J.* **50**: 424–30.

Shekhar, V. C., Iritani, W. M. & Arteca, R. (1978). Changes in ascorbic acid content during growth and short term storage of potato tubers (*Solanum tuberosum* L.). *Am. Potato J.* **55**: 663–70.

Singh, M. & Verma, S. C. (1979). Post-harvest technology and utilization of potato. In H. Kishore (ed.), *Post-harvest technology and utilization of potato*. International Potato Center, Region VI, New Delhi.

Smith, O. (1975). Potato chips. In W. F. Talburt & O. Smith (eds.), *Potato processing*, 3rd edn. AVI Publishing Company, Inc., Westport, CT.

Smith, O. (1977). *Potatoes: production, storing, processing*, 2nd edn. AVI Publishing Co. Inc., Westport, CT.

Steele, L., Jadhav, S. & Hadziyev, D. (1976). The chemical assay of vitamin C in dehydrated mashed potatoes. *Lebensm.-Wiss. Technol.* **9**: 239–45.

Strachan, P. W. (1983). Optimised quality in frozen foods. *Inst. Food Sci. Technol. Proc.* **16**: 65–74.

Streightoff, F., Munsell, H. E., Ben-Dor, B., Orr, M. L., Cailleau, R., Leonard, M. H., Ezekiel, S. R., Kornblum, R. & Koch, K. G. (1946). Effect of large-scale methods of preparation on vitamin content of food. I. Potatoes: *J. Am. Diet. Assoc.* **22**: 117–27.

Swaminathan, K. & Gangwar, B. M. L. (1961). Cooking losses of vitamin C in Indian potato varieties. *Indian Potato J.* **3**: 86–91.

Sweeney, J. P., Hepner, P. A. & Libeck, S. Y. (1969). Organic acid, amino acid and ascorbic acid content of potatoes as affected by storage conditions. *Am. Potato J.* **46**: 463–9.

Szkilladziowa, W., Secomska, B., Nadolna, I., Trzebska-Jeska, I., Wartanowicz, M. & Rakowska, M. (1977). Results of studies on nutrient content in selected varieties of edible potatoes. *Acta Aliment. Pol.* **3**: 87–97.

Talburt, W. F. (1975a). History of potato processing. In W. F. Talburt & O. Smith (eds.), *Potato processing*, 3rd edn. AVI Publishing Company, Inc., Westport, CT.

Talburt, W. F. (1975b). Canned white potatoes. In W. F. Talburt & O. Smith (eds.), *Potato processing*, 3rd edn. AVI Publishing Company, Inc., Westport, CT.

Talley, E. A., Fitzpatrick, T. J. & Porter, W. L. (1964). Chemical composition of potatoes. IV. Relationship of the free amino acid concentrations to specific gravity and storage time. *Am. Potato J.* **41**: 357–66.

Talley, E. A. & Porter, W. L. (1970). Chemical composition of potatoes. VII. Relationship of the free amino acid concentrations to specific gravity and storage time. *Am. Potato J.* **47**: 214–24.

Talley, E. A., Toma, R. B. & Orr, P. H. (1983). Composition of raw and cooked potato peel and flesh: Amino acid content. *J. Food Sci.* **48**: 1360–1, 1363.

Talley, E. A., Toma, R. B. & Orr, P. H. (1984). Amino acid composition of freshly harvested and stored potatoes. *Am. Potato J.* **61**: 267–79.

Thomas, P. (1984). Radiation preservation of foods of plant origin. Part 1. Potatoes and other tuber crops. *CRC Crit. Rev. Food Sci. Nutr.* **19**: 327–79.

Toma, R. B., Augustin, J., Orr, P., True, R. H., Hogan, J. M. & Shaw, R. L. (1978a). Changes in the nutrient composition of potatoes during home preparation. I. Proximate composition. *Am. Potato J.* **55**: 639–45.

Toma, R. B., Augustin, J., Shaw, R. L., True, R. H. & Hogan, J. M. (1978b). Proximate composition of freshly harvested and stored potatoes. *J. Food Sci.* **43**: 1702–4.

Treadway, R. H., Heisler, E. G., Whittenberger, R. T., Highlands, M. E. & Getchell, J. S. (1955). Natural dehydration of cull potatoes by alternate freezing and thawing. *Am. Potato J.* **32**: 293–303.

True, R. H., Hogan, J. M., Augustin, J., Johnson, S. R., Teitzel, C., Toma, R. B. & Orr, P. (1979). Changes in the nutrient composition of potatoes during home preparation. III. Minerals. *Am. Potato J.* **56**: 339–50.

Walker, R. & El-Harith, E. A. (1978). Nutritional and toxicological properties of some raw and modified starches. *Ann. Nutr. Aliment.* **32**: 671–9.

Watt, B. K. & Merrill, A. L. (1975). *Composition of foods: raw, processed, prepared*. Agriculture Handbook no. 8. US Dept. of Agriculture, Washington, DC.

Weaver, M. L., Reeve, R. M. & Kueneman, R. W. (1975). Frozen french fries and other frozen potato products. In W. F. Talburt & O. Smith (eds.), *Potato processing*, 3rd edn. AVI Publishing Company, Inc., Westport, CT.

Weaver, M. L., Ng, K. C. & Huxsoll, C. C. (1979). Sampling potato tubers to determine peel loss. *Am. Potato J.* **56**: 217–24.

Weaver, M. L., Timm, H. & Ng, H. (1983). Changes in nutritional composition of Russet Burbank potatoes by different processing methods. *Am. Potato J.* **60**: 735–44.

Weaver, M. L., Timm, H., Nonaka, M., Sayre, R. N., Ng, K. C. & Whitehand, L. C. (1978). Potato composition. III. Tissue selection and its effects on total nitrogen, free amino acid nitrogen and enzyme activity (polyphenolase, monophenolase, peroxidase, and catalase). *Am. Potato J.* **55**: 319–31.

Werge, R. W. (1979). Potato processing in the central highlands of Peru. *Ecol. Food Nutr.* **7**: 229–34.

Werner, H. O. & Leverton, R. M. (1946). The ascorbic acid content of Nebraska-grown potatoes as influenced by variety, environment, maturity and storage. *Am. Potato J.* **23**: 265–7.

Willard, M. (1975). Potato flour. In W. F. Talburt & O. Smith (eds.), *Potato processing*, 3rd edn. AVI Publishing Company, Inc. Westport, CT.

Willard, M. & Kluge, G. (1975). Potato flakes. In W. F. Talburt & O. Smith (eds.), *Potato processing*, 3rd edn. AVI Publishing Company, Inc., Westport, CT.

Wills, R. B. H., Wimalasiri, P. & Greenfield, H. (1984). Dehydroascorbic acid levels in fresh fruit and vegetables in relation to total vitamin C activity. *J. Agric. Food Chem.* **32**: 836–8.

Witkowski, C. & Paradowski, A. (1975). [Effect of the time and temperature of sterilisation on vitamin C in canned potatoes.] In Polish. *Przem. Ferment. Rolny* **19**: 7–8.

Yamaguchi, M., Perdue, J. W. & MacGillivray, J. H. (1960). Nutrient composition of White Rose potatoes during growth and after storage. *Am. Potato J.* **37**: 73–6.

Young, N. A. (1981). *Potato products: production and markets in the European Communities*, Commission of the European Communities Information on Agriculture no. 75. Office for Official Publications of the European Communities, Luxembourg.

Zarneger, L. & Bender, A. E. (1971). The stability of vitamin C in machine-peeled potatoes. *Proc. Nutr. Soc.* **30**: 94a.

Zobel, M. (1979). [Nutritional aspects of potato peeling in the DDR and from the international viewpoint.] In German. *Ernaehrungsforschung* **24**: 74–81.

5

Glycoalkaloids, proteinase inhibitors and lectins

Although a valuable food, the potato has toxic, or potentially toxic, constituents: glycoalkaloids, proteinase inhibitors and lectins. These have been the subject of much research and debate, particularly in recent years. In this chapter, each toxic component group is reviewed, and its structure and probable function within the general physiology of the potato plant described briefly. Emphasis, however, is on the current consensus of opinions regarding nutritional and physiological significance of these components for human beings (for other reviews, see Jadhav & Salunkhe, 1975; Maga, 1980; Morris & Lee, 1984).

Glycoalkaloids
Chemical structure and content in the tuber

The Solanaceae family is recognized for the numerous alkaloids found among its various member species. Alkaloids are nitrogen-containing organic compounds occurring in plants, as well as in a small number of animal products (Robinson, 1974). As a result of the diverse pharmacological properties of alkaloids, many plants have long been used as drug sources; some were prescribed for their curative or beneficial effects; many others have become well known for their poisonous, aphrodisiac, narcotic or hallucinogenic attributes.

Two such narcotic and hallucinogenic plants, the mandrake and deadly nightshade, are related to the potato. When first introduced into Europe, potatoes may have been shunned because of their 'guilty association' with such notorious relatives (Rhoades, 1982). Under normal conditions of human consumption, the amounts of potato alkaloids ingested are not harmful. Sometimes, however, alkaloid quantities can increase to toxic, and in rare instances fatal, levels. It is necessary, therefore, to understand how such toxic levels arise and what can be done to prevent them.

Plants of the genus *Solanum* contain carbohydrate derivatives of 3-hydroxysteroidal alkaloids (Osman & Sinden, 1977), collectively referred to as glycoalkaloids. More simply, glycoalkaloids are alkaloids with one or more sugar residues attached (Wood & Young, 1974). They are weakly basic substances, readily soluble in weak acids and acidified alcohols and only slightly soluble in water (Zitnak, 1964). The major glycoalkaloids in most cultivated potato species are α-solanine and α-chaconine, both of which are derived from the alkaloid aglycone solanidine (Maga, 1980). Table 5.1 shows the alkaloids, and their glycosides, of the major tuber-bearing *Solanum* spp. A new glycoalkaloid, commersonine, has been identified in the wild tuber-bearing *S. commersonii* (Osman et al., 1976).

Glycoalkaloids are found in most tissues of the potato, except in the pith or centre portion of the tuber (Maga, 1980). Highest levels are usually found in sprouts, flowers or other actively growing areas

Table 5.1. *Glycoalkaloids of tuber-bearing* Solanum *spp.*[a]

Aglycone	Glycoalkaloid	Carbohydrate component[b]
Solanidine	α-Solanine[c]	Solatriose
	α-Chaconine[c]	Chacotriose
	Dehydrocommersonine	Commertetrose
Demissidine	Demissine	Lycotetraose
	Commersonine	Commertetrose
Acetylleptinidine	Leptine I[d]	Chacotriose
	Leptine II[d]	Solatriose
Tomatidenol	α-Solamarine	Solatriose
	β-Solamarine	Chacotriose
Tomatidine	α-Tomatine	Lycotetraose

[a] Table constructed by T. Johns, personal communication.
[b] Solatriose   D-galactose — L-rhamnose / D-glucose
   Chacotriose   D-glucose — L-rhamnose / L-rhamnose
   Commertetrose   D-galactose — D-glucose — D-glucose / D-glucose
   Lycotetraose   D-galactose — D-glucose — D-glucose / D-xylose
[c] β- and γ-solanines and β- and γ-chaconines are products of a partial hydrolysis of the respective α-glycosides.
[d] The action of esterases or mild alkaline treatment produces leptines I and II.

(Table 5.2). Concentrations are higher in immature potatoes and are diluted as the tuber enlarges. In normal tubers, most glycoalkaloids (60% to 80%) are concentrated in the outer layers (see Table 5.3) and may be removed with the peel (Wood & Young, 1974; Bushway *et al.*, 1983). Prolonged storage, however, causes glycoalkaloids to migrate toward the centre of the tuber. Bitter potatoes have high glycoalkaloid levels throughout the flesh (Table 5.3), so peeling may remove only 30% to 35% of the total amount present (Wood & Young, 1974).

The total quantity of glycoalkaloids in a potato tuber varies greatly according to species and variety. Table 5.4 shows the glycoalkaloid contents of some common American and European commercial varieties. It is important to note that glycoalkaloid levels in the majority of these varieties are below 10 mg/100 g (FWB) and are therefore imperceptible by taste.

Potato tubers with more than 20 mg glycoalkaloids/100 g (FWB) are generally considered to be beyond the upper safety limit for consumption purposes (Jadhav & Salunkhe, 1975). Ross *et al.* (1978), however, suggest that the acceptable limit should be much lower than this (6 to 7 mg/100 g), as large glycoalkaloid level variations are produced within a single cultivar by differences in locality and season. This is still open to debate.

Levels higher than 20 mg/100 g (FWB) are found among wild *Solanum* and highland cultivated species (Table 5.5), which are used extensively in breeding programmes to obtain characteristics such as disease resistance and cold hardiness (Osman *et al.*, 1978). Sanford & Sinden (1972) concluded that glycoalkaloid content was highly heritable. Additionally, many of these species contain considerable quantities of glycoalkaloids not found in currently cultivated commercial varieties (Table 5.6).

Several authors (Osman *et al.*, 1976; Maga, 1980; Jadhav & Salunke, 1975; Zitnak & Johnston, 1970; Gregory *et al.*, 1981) noted that the

Table 5.2. *Distribution of glycoalkaloids in the potato plant*

Plant part	Glycoalkaloids (mg/100 g (FWB))
Sprouts	200–400
Flowers	300–500
Stems	3
Leaves	40–100

From Wood & Young (1974).

incorporation of wild *Solanum* spp. in a breeding programme may produce tubers with higher glycoalkaloid contents and/or introduce glycoalkaloids, other than solanine and chaconine, thus creating a potential health hazard for humans. How the presence or absence of solanine, chaconine and commersonine is inherited in the accessions of a wild species, *S. chacoense*, is explained by McCollum & Sinden (1979).

According to Osman *et al.* (1978), the glycoalkaloids listed in Table 5.6 are of the same order of toxicity as α-solanine and α-chaconine, and that they present no greater hazard than glycoalkaloids now found in commercial varieties. However, most plant breeders try to produce varieties with low levels of glycoalkaloids. Wild species to be used as parents should be screened for abnormal glycoalkaloid concentrations.

Maga (1980) commented on the effects of cooking on glycoalkaloids. It is generally thought that cooking does not cause changes: Baker *et al.* (1955) reported little or no loss of solanine on either cooking or freeze-drying. Maga recommended investigations into whether cooking reduces the levels or simply enhances the difficulty of glycoalkaloid extraction. Since Zitnak & Johnston (1970) showed that the initial decomposition temperature of solanine is 243 °C and the melting point is 285 °C, it would seem unlikely that potato glycoalkaloids would be destroyed by home preparations or by most types of commercial processing. Bushway & Ponnampalam (1981) found that both α-chaconine and α-solanine were stable to boiling and baking and, of the processes they tested, only frying reduced glycoalkaloid content. Significant decreases (DWB) in the total

Table 5.3. *Normal glycoalkaloid levels in various tuber tissues*

Tuber tissue	Glycoalkaloids (mg/100 g (FWB))
Normal tuber	
Skin (2% to 3% of tuber)	30–60
Peel and eye (3 mm disc around eye)	30–50
Peel (10% to 15% of tuber)	15–30
Whole tuber	7.5
Flesh	1.2–5
Bitter tuber	
Peel	150–220
Whole tuber	25–80

Based on data by Wood & Young (1974).

Table 5.4. *Glycoalkaloid (solanine and chaconine) contents of some common North American and European commercial potato varieties*[a]

Variety	Glycoalkaloid content (mg/100 g (FWB))
'Columbia russet'	1.8
'Russet rural'	2.1
'Sebago'	2.1
'McCormick'	2.6
'Jubel'	3.2
'Triumph'	3.6
'King'	3.7
'Brown beauty'	3.7
'Earlaine'	3.9
'Green mountain'	4.0
'Wee McGregor'	4.2
'Jossing'[b]	4.2
'Red Pontiac'	4.3
'Russet Burbank'[c]	4.7
'Early Ohio'	4.8
'Mesaba'	5.2
'Hundred-day cobbler'	5.3
'Arnica'	5.4
'Katahdin'	5.5
'Irish cobbler'	5.6
'Ackersegen'	5.8
'Cobbler'	6.4
'Houma'	6.4
'Chippewa'	6.6
'Libertas'[b]	6.8
'Warba'	6.9
'Blue salad'	7.0
'Early rose'	7.2
'Netted Gem'[d]	7.9
'Golden'	8.8
'Early epicure'	8.9
'Furore'[b]	9.2
'White-blossomed cobbler'	9.4
'As'[b]	9.6
'Gineke'[b]	9.6
'Kennebec'[d]	9.7
'Pioneer rural'	10.0
'Erstling'	10.0
'Hindenger'	11.6
'Rural New Yorker'	13.0
'Kerr's pink'[b]	13.6
'Pimpernell'[b]	15.3
'King George'[b]	18.7
'Prestkvern'[b]	34.5

Table 5.5. *Total glycoalkaloid (α-solanine and α-chaconine) content of selected species of* Solanum *tubers*[a]

Species	(mg/100 g (FWB))
S. tuberosum	6.4[b]
S. tuberosum group andigena	4.9
S. acaule	79.8
	126
	53
	35
	58.5
S. ajanhuiri	7.1
S. curtilobum	29.0
	21.8
	18.3
	3.8
S. juzepczukii	15.9
	11.7
	19.2
	41.3
	18.8
	46.8
S. stenotonum	3.6
	6.1

[a] Figures from Osman et al. (1978). Reprinted with permission from Osman et al., *J. Agric. Food Chem.* **26**: 1246–8. Copyright 1978, American Chemical Society.
[b] Each value represents the mean of duplicate analyses of freeze-dried tuber extracts. *S. acaule* is a wild species; the rest are cultivated.

Notes to Table 5.4.

[a] Adapted from Maga (1980).
[b] European varieties.
[c] W. G. Burton (personal communication, 1983) stated that 'Netted Gem' is a synonym for 'Russet Burbank'. He noted that the two have different glycoalkaloid values, and that different maturity levels or growing conditions could cause these differences. No such information is available for the samples, so these issues are open to speculation. Burton's comments highlight, however, the important effects that environment and physical state of the tuber can have on glycoalkaloid levels.
[d] From Wood & Young (1974).

Table 5.6. *Glycoalkaloids of selected* Solanum *spp.*[a]

Species	α-Solanine	α-Chaconine	β-Chaconine	Solamarines[b]	Demissine	Tomatine
S. ajanhuiri[c]	57.3[d]	39.0	3.5	—	—	—
S. curtilobum	46.4	34.8	—	5.3	13.4	—
S. stenotonum	24.7	69.8	5.5	—	—	—
S. juzepczukii	37.8	14.0	—	7.7	40.4	—
S. acaule[e]						
1	—	—	—	—	95.5	—
2	—	—	—	—	62.1	30.9
3	—	—	—	—	88.2	11.6
4	—	—	—	—	64	34

[a] Figures from Osman *et al.* (1978). Reprinted with permission from Osman *et al.*, *J. Agric. Food Chem.* **26**: 1246–8. Copyright 1978, American Chemical Society.
[b] Combined value for α- and β-solamarine.
[c] All species are cultivated except *S. acaule*.
[d] Values represent percentage of total glycoalkaloids.
[e] Four clones of *S. acaule* were analysed.

glycoalkaloid content of the cortex tissues of three cultivars occurred on baking and frying (Ponnampalam & Mondy, 1983). However, the authors suggest that the decrease in total glycoalkaloids may have been due to the difficulty of extracting them from cooked tissues. In spite of the decrease, glycoalkaloids were concentrated greatly by moisture loss during baking and frying of cortex (to produce fried potato peels, see Chapter 4). This resulted in levels greatly in excess of the recommended safe concentration in fried peels from two of the three cultivars (43 mg/100 g and 65 mg/100 g (FWB)). The analysis of commercially produced fried peels by Bushway & Ponnampalam (1981) revealed levels of 139 to 145 mg of combined α-chaconine and α-solanine per 100 g of product. Both groups of workers recommended caution in eating these peels. Sizer et al. (1980) found that total alkaloids were not destroyed by frying to produce potato chips and that concentrations increased during processing, due to water loss. The only effective way to control these alkaloids is to continue to grow and develop varieties inherently low in average glycoalkaloid content (Sinden & Webb, 1974), and to inhibit their synthesis in the tubers (Jadhav & Salunkhe, 1975).

In the late 1960s, the variety 'Lenape', a descendant of *S. chacoense*, was removed from commerce by the US Department of Agriculture and the Canadian Department of Agriculture because of its high glycoalkaloid content, which ranged from 18.6 to 35.4 mg/100 g (FWB), depending upon location (Jadhav & Salunkhe, 1975; Zitnak & Johnston, 1970). 'Lenape' was considered to be a promising variety because of its immunity to virus A, resistance to common races of late blight, high specific gravity, low sugar content, and excellent chipping qualities (Zitnak & Johnston, 1970). Although 'Lenape' is an isolated case of a released variety that had to be withdrawn due to excessive glycoalkaloid content, similar situations are foreseeable, especially in light of current widespread efforts to breed greater resistance and climatic tolerances into potatoes.

Physiological functions

Glycoalkaloids are believed to be part of the disease and pest resistance mechanisms of potato plants and tubers. For example, α-solanine and α-chaconine were highly toxic to the fungus *Helminthosporium carbonum* (Allen & Kuć, 1968), and the growth of early blight (*Alternaria solani*) on potato dextrose agar is inhibited by potato alkaloids (Sinden et al., 1973). Tingey et al. (1978) found a significant correlation between foliar glycoalkaloid levels and resistance to potato leaf hopper, *Empoasca fabae*; however, they believed other natural toxicants and nutritional factors might also have contributed to the result. Other researchers have noted resistance to Colorado beetle in *Solanum demissum* and *S. chacoense* due to the glycoalkaloids demissine and leptin (Schwarze, 1962).

The depressive action of foliar glycoalkaloids on certain insects suggests potential for glycoalkaloid-based breeding for resistance to these pests. However, levels of tuber and foliar glycoalkaloids are highly correlated and the frequency of clones with high foliar level contents but safe tuber level contents is extremely low in segregating populations (Raman *et al.*, 1979). If simple, rapid screening methods were used, and if individual glycoalkaloids could be identified as deterrents to insect pests but less toxic to humans or animals, then glycoalkaloid-based resistance might be exploited to a greater extent. In this connection, the accuracy of six extraction methods for samples used to determine glycoalkaloid content has been studied by Smittle (1971), and Bergers (1980) described a rapid means of assaying solanidine glycoalkaloids. Also, new rapid methods of analysis suitable for plant breeders have been developed (Coxon *et al.*, 1979; Coxon & Jones, 1981; Morris & Lee, 1981; Ross *et al.*, 1978; Morgan *et al.*, 1983, 1985).

In 1972, Renwick proposed that potatoes resistant to late blight (*Phytophthora infestans*) contained higher levels of glycoalkaloids than did non-resistant varieties. However, a subsequent test comparing clones with late blight resistance to susceptible clones found that, in both clones, blight infection did not cause higher glycoalkaloid levels, nor were glycoalkaloid contents of tubers from late blight inoculated plants any higher than those of tubers from healthy (fungicide-protected) plants (Deahl *et al.*, 1973).

Most researchers currently agree that, although glycoalkaloids are not directly responsible for late blight resistance, they function, in part, as members of the group called *phytoalexins*, which include all compounds formed before or after infection by a pathogen that contribute to the resistance of infected tissue (Kuć, 1972). In plants resistant to the pathogen, an elicitor induces these compounds to accumulate as part of a hypersensitive reaction which thus prevents the pathogen from spreading throughout the plant or tuber (Henfling, 1979). Relatively little work has been done on the human nutritional effects of breeding potatoes with greater disease resistance.

Effect on potato flavour

Potato tuber glycoalkaloids significantly influence flavour. In concentrations of less than 10 mg/100 g (FWB), they are normally imperceptible by taste, though individual people have varying thresholds of perception. Glycoalkaloid levels above 10 mg/100 g (FWB) usually impart to the tuber a bitter taste which develops within 15 to 30 s (Wood & Young, 1974). This taste is described as a slowly developing, hot/burning, persistent irritation at the sides of the tongue and the back of the mouth, not unlike the sensation imparted by hot peppers. Tubers with glycoalkaloid contents of 20 mg/100 g or greater give an immediate burning sensation and are generally considered unfit for human consumption. The highland Andean people, however, have found ways round the problem (see later in this chapter).

Some workers have proposed that phenols are more likely to be responsible for bitterness than glycoalkaloids; however, Sinden et al. (1976) showed that glycoalkaloid levels, not those of phenols, correlated with taste perceptions of burning and bitterness in tubers (Table 5.7). In addition to these sensations, the taste panelists noted that tubers with medium to high glycoalkaloid levels had a 'metallic aftertaste', 'a bite', caused 'a coating of the teeth and mouth' and generally imparted an

Table 5.7. *Effects of glycoalkaloid and phenolic contents of potatoes on sensory ratings of bitterness and of burning*[a]

Clone[b]	Glycoalkaloid content (mg/100 g)	Phenolic content (mg/100 g)	Bitterness[c] content (0–4 scale)	Burning[c] content (0–4 scale)
1	58.0	29	2.4	3.4
2	51.0	43	2.2	3.2
3	25.0	17	1.8	2.0
4	23.0	23	0.9	1.7
5	22.0	41	1.3	1.7
6	22.0	33	1.9	1.7
7	14.0	59	0.8	0.6
8	7.3	27	0.0	0.1
9	5.9	30	0.2	0.2
10	4.4	31	0.1	0.0
11	2.0	29	0.1	0.1
12	0.9	24	0.1	0.0
13	0.7	21	0.0	0.0
LSD (0.05)	5.7	6.7	0.64	0.71

LSD, least significant difference.
[a] From Sinden et al. (1976). Reprinted from *J. Food Sci.* 1976, **41** (3): 521. Copyright © by the Institute of Food Technologists.
[b] Clones 1 to 8 are breeding lines; 9 to 13 are cultivars.
[c] Means of 18 evaluations; 0 = no bitterness or burning, 4 = very strong bitterness or burning.

'unpleasant aftertaste' (Sinden et al., 1976). It is important to note, however, that many of these 'negative tastes' can be masked in potato preparations by using hot sauces, spices, gravies or oils, and people can and do unknowingly ingest sufficient quantities of glycoalkaloids to cause toxic reactions.

Accumulation

Under certain conditions, glycoalkaloid concentrations can increase to toxic levels in the tuber. As well as the genetic varietal differences in glycoalkaloid content, external conditions can cause increases in quantities of glycoalkaloids, varieties with an inherently high average glycoalkaloid content being particularly susceptible to excessive glycoalkaloid production.

Greening

Exposure to light causes potatoes to turn green due to the production of chlorophyll. Many people commonly associate the greening of potato tubers with increases in bitterness, caused by rising glycoalkaloid levels. Maga (1980) states, however, that although exposure to light induces both greening and increased glycoalkaloid levels, the two processes are independent. Gull & Isenberg (1960) demonstrated that total solanine content and susceptibility to greening in potatoes are varietal characteristics. In their experiments with light exposure, 'Katahdin', the variety which developed the highest concentration of solanine, was one of the least susceptible to greening, while 'Cherokee', which greened most readily, developed only moderate increases in solanine content. They concluded that the amount of greening is probably not a reliable indicator of glycoalkaloid level in potato tubers. Nair et al. (1981) showed that, contrary to the findings of earlier studies, there is a close relationship between chlorophyll and solanine synthesis. This topic needs further investigation with respect to disease and insect control.

Reviewing the effects of irradiation (by exposure to gamma rays) on the production of chlorophyll and solanine in tubers, Thomas (1984) quoted most researchers as having found that irradiation delayed or inhibited greening in tubers subsequently exposed to light. However, some researchers observed inhibition of glycoalkaloid synthesis with irradiation and others no effect.

Manifest greening of potatoes is a general indication that undesirable levels of glycoalkaloids may be present (Jadhav & Salunkhe, 1975). Therefore greening during marketing represents a high economic loss due to rejection of greened potatoes as unfit for consumption (Morris & Lee, 1984).

Glycoalkaloids 173

Effects of varying light exposure

Duration, intensity and colour of light exposure have different effects on glycoalkaloid content (Figure 5.1). Longer day-length (18 h of light) causes greater tuber solanine production than shorter (10 h) day-length (Wolf & Duggar, 1946). Maga (1980), in reviewing previous

Figure 5.1. Post-harvest methods of handling potatoes can affect glycoalkaloid levels. Storing potatoes in the house until time for sale or consumption in Bangladesh helps to maintain normal glycoalkaloid levels (above). Potatoes left in the field and exposed to direct sunlight during sorting in Tunisia may increase glycoalkaloid levels (below).

research, noted that high light intensity and greater duration of exposure to light resulted in increased glycoalkaloid levels, while all types of coloured light produced less glycoalkaloid increase over time than did normal daylight.

Baerug (1962) studied the amount of light to which tubers can be exposed without a serious increase in solanine, assuming that some exposure is unavoidable during harvesting and transportion. When freshly dug tubers were exposed to bright sunlight for 2 to 4 h, only minor changes in solanine content occurred. However, there was a significant increase (5 to 20 mg/100 g (FWB)) with an exposure of 6 h. Cooking quality, as judged by flavour and surface discoloration, deteriorated seriously with exposure to light for more than 6 h. Solanine increases were greatest in the peel, moderate in the cambium layer (outer 2 to 10 mm of flesh) and insignificant in the remainder of the tuber. Washed tubers were more susceptible to light than brushed ones.

Environmental effects

Few studies have dealt with environmental effects on glycoalkaloid content. Ross *et al.* (1978) found significant variations in glycoalkaloid levels in varieties grown at different localities in Germany. They also found year-to-year differences in the same variety grown in the same location. Sikilinda & Kiryukhin (1975) reported that the accumulation of glycoalkaloids in the tubers of 19 potato varieties in the Moscow region differed between cultivars and depended on growing conditions.

In reviewing glycoalkaloids, Jadhav & Salunkhe (1975) concluded that unfavourable conditions, such as nutritional imbalance, frost or hail damage before tuber maturity, an unusually cool growing season and a high number of overcast days, could cause a build-up of glycoalkaloids. They noted that soil conditions in different localities played a minor role and that reports on the effects of moisture and organic content of soils were conflicting. An earlier study (Arutyunyan, 1940) claimed that the glycoalkaloid content of potatoes grown in mountainous regions is always less than that of those grown in hot climates.

In a study of environmental effects on glycoalkaloid content, Sinden & Webb (1974) stated that it was not possible to determine, with certainty, the environmental factors or cultivation practices that affect glycoalkaloid content in potatoes. Although tuber damage from pests, wounding or bruising and any conditions which delayed tuber growth or maturation resulted in higher glycoalkaloid levels, the investigators found that varietal differences were more important in determining glycoalkaloid content. However, varieties with high average glycoalkaloid content are more likely to produce excessive glycoalkaloid concentrations when subjected to poor environmental conditions or improper handling.

Wounding

Glycoalkaloids may also be synthesized in potatoes as a result of wounding (e.g. bruising, cutting or slicing) during harvesting or post-harvest handling. Fitzpatrick *et al.* (1978) found that 10% (or more) of the loose white potatoes available in retail markets (in an eastern state of the USA) had some degree of mechanical damage. Analysis of three locally available commercial varieties, 'Katahdin', 'Russet Burbank' and 'Red Pontiac', revealed that damaged areas (cracks, fissures, bruising) had higher glycoalkaloid levels than undamaged ones, but in no case did the concentration exceed 10 mg/100 g (FWB). However, 8 of the 12 'Katahdin' samples had overall levels higher than 10 mg/100 g (FWB) and 5 of these were over 14 mg. This evidence, in their opinion, did not rule out damage as a cause of toxic levels (>20 mg/100 g (FWB)), but only that it did not happen in most of the varieties they examined.

Wu & Salunkhe's (1976) study of mechanical injuries showed that the extent of glycoalkaloid formation depends on cultivar, type of mechanical injury, storage temperature and duration of storage. Most accumulation occurred during the first 15 days following injury, with little increase noted after 30 days. [It should be noted that severe damage occurring in the tests was not necessarily comparable with normal handling of potatoes.] Ahmed & Müller (1978) found that damage, storage conditions (light) and period of storage significantly affect potato glycoalkaloid content. Amounts of solanine and chaconine increased after damage and then further increased when tubers were exposed to light during storage. Chips showed nearly the same trend as raw potatoes. These conclusions concur with those of Fitzpatrick *et al.* (1978) that mechanical injury can elevate glycoalkaloid levels, and indicate that care should be taken to avoid damage to tubers.

Effects of storage

Storage, and particularly the temperature during this period, can also cause glycoalkaloid build-up. Studies have generally shown that low temperature storage causes or maintains more bitterness in potatoes than do temperatures above 10 °C (Jadhav & Salunkhe, 1975). Bushway *et al.* (1981) reported higher glycoalkaloid levels during storage at 3.3 °C than at 7.7 °C in North American varieties. In contrast, Linnemann *et al.* (1985) reported a lower content of glycoalkaloids in a European variety ('Bintje') stored at 7 °C than in those at 16 °C or 28 °C after 12 weeks, but decreases at all three temperatures during storage were slight, perhaps because of the low initial content of 3.6 mg/100 g (FWB). Fitzpatrick *et al.* (1977) found that at 7 °C and 85% rel. hum. there was little increase in glycoalkaloid content, but, when potatoes were peeled, sliced or otherwise wounded and then held at room temperature for an extended period without blanching or cooking, glycoalkaloids, especially solanine, accumulated to undesirable levels. The accumulation as a result of tuber injury is part of a 'wound response', which is thought to be associated with the tuber's defence mechanisms (Salunkhe *et al.*, 1972).

Maga (1981) measured glycoalkaloid accumulation in sliced potatoes under different pre-processing storage conditions, and observed significant increases after only 1 h of storage at 5 °C and at 25 °C: glycoalkaloids increased more at higher temperatures. With 7 h of storage, total glycoalkaloids had more than doubled in slices that were not soaked in water. Slices that were soaked in water also showed increases but had lower glycoalkaloid levels at each time point than the unsoaked slices. Results from this study show that relatively short storage of unpeeled, sliced potatoes can result in increased glycoalkaloid levels. To avoid accumulation of high levels of glycoalkaloids in potato products, processing should start immediately after peeling (Salunkhe *et al.*, 1972).

Processing effects

Processing methods can affect the glycoalkaloid contents of potato products. Maga (1981) notes that potential glycoalkaloid build-up becomes especially important in making chips (crisps) because the process removes water and thus concentrates the total glycoalkaloids. Sizer *et al.* (1980) also found that potato chipping did not destroy glycoalkaloids and resulted in increased total concentration. They showed that removal of the peel from potato slices significantly lowered the total glycoalkaloid content of finished chips. Two of three samples of commercial chips from the USA contained 9.5 mg, and the third 72 mg, of glycoalkaloids per 100 g of chips. The high glycoalkaloid sample still retained a considerable amount of peel. The same authors also point out that salt and oil can mask the bitter taste caused by even high concentrations of glycoalkaloids, as would other flavourings now added to potato chips.

When Bushway & Ponnampalam (1981) compared several methods of potato preparation (baking, boiling, microwave and deep-fat frying), only frying reduced the total glycoalkaloid content significantly.

Zaletskaya *et al.* (1977) showed that processing of potatoes into dried purée lowered glycoalkaloid levels. They reported that fresh 'Loshitskii' tubers contained 31 mg/100 g (FWB). After mechanical cleaning and boiling, this level dropped to 7 mg/100 g. Drying reduced it further to 5.25 mg; only 3.55 mg/100 g remained after 12 months of storage.

Andean people living in the highlands of southern Peru and Bolivia have long recognized the benefits of processing in the utilization of bitter potatoes (Antúnez de Mayolo, 1981; Christiansen & Thompson, 1976; Werge, 1979). Bitter potatoes with high levels of glycoalkaloids (over 30 mg/100 g (FWB)) are grown throughout the highest zones of the Andes (>3600 m). They are resistant to frosts common at those altitudes, but their high glycoalkaloid content makes them generally inedible as fresh food, although they are sometimes eaten fresh in the Bolivian Altiplano (Johns, personal communication, 1983).

Several methods of processing bitter potato have been developed over the centuries and result in products known locally as *chuño blanco* (or *tunta* in southern Peru and Bolivia), *chuño negro* and *tokosh* (see Chapter 4). The traditional processing techniques remove the glycoalkaloids, leaving products with less than 10 mg/100 g (FWB) (Christiansen & Thompson, 1976). The advantage of this kind of processing is that the products can be stored for long periods without deterioration: *chuño* has been found in perfect condition in 500-year-old coastal Inca tombs.

Traditional potato processing is also practised when there are surpluses of normal non-bitter potatoes. These are cooked and solar-dried to make *papa seca* or dried potatoes (see Chapter 4). Only intact, whole potatoes are used in this type of processing, which does not involve the extraction of glycoalkaloids. It has been proposed that techniques such as those for making *papa seca* could also be used to improve the marketability of potato tubers that are either too small or too damaged for local consumer retail markets (Shaw & Booth, 1982). However, due to the 'wound response', the build-up of glycoalkaloids during light exposure or poor storage conditions, and the concentration of glycoalkaloids in small or immature tubers, it is possible that subsequent processing of damaged, green, small or diseased potatoes could result in products with unacceptably high glycoalkaloid levels. Christiansen (1977), however, showed that the *papa seca* process reduces the glycoalkaloid level of bitter potatoes from about 30 mg/100 g to about 6 mg/100 g.

Toxicity

Accidental consumption of potatoes containing high levels of glycoalkaloids has caused severe illness and, on rare occasions, death (Jadhav & Salunkhe, 1975). There have been at least 12 separate documented instances of glycoalkaloid poisoning involving up to 2000 people, with about 30 deaths (Morris & Lee, 1984). Acute illness caused by glycoalkaloids is perhaps more prevalent than indicated by the medical records because the symptoms are common to many ailments and may be easily mistaken for severe digestive discomfort (gastroenteritis) with nausea, diarrhoea, vomiting, stomach cramps, headaches, and dizziness (Wood & Young, 1974).

Jadhav & Salunkhe (1975) documented much of the early history relating to human glycoalkaloid poisoning and toxicity. More recently, Morris & Lee (1984) reviewed the mechanisms of toxicity. Glycoalkaloids may have two toxic actions: inhibition of cholinesterase, thus affecting the nervous system, and disruption and injury to membranes in the gastrointestinal tract and elsewhere. A surge in research on glycoalkaloid toxicity occurred in 1972 when Renwick (1972) proposed that unknown factors (possibly glycoalkaloids) associated with severity of late blight and resistance to this disease in potatoes were correlated geographically with the incidence of anencephaly and spina bifida cystica (ASB) in humans. He further recommended that women likely to bear children should avoid consuming imperfect potatoes (Renwick, 1973).

Subsequent failure to correlate a specific teratogen from potatoes with the occurrence of ASB, caused Renwick's hypothesis to be abandoned (Anon., 1975).

Potatoes with high glycoalkaloid levels were recently shown to cause anaemic and hyperglycaemic conditions in rabbits (Ahmad, 1982). When fed to pregnant does at a rate of 208 g potato/kg rabbit body weight per day (30.0 mg glycoalkaloids/100 g potato (FWB)), the high-glycoalkaloid potatoes caused litter mortality of 55.6% compared to 15.7% with normal potatoes (8.78 mg glycoalkaloids/100 g (FWB)). It was suggested that glycoalkaloids may prevent lactation, causing death by starvation in newly born infants. They may also enter foetal circulation and depress foetal breathing: 23% of the rabbit litter died prior to delivery. Pierro *et al.* (1977) showed, although with limited data, that under certain experimental conditions, purified α-chaconine exerts harmful effects on the developing central nervous system of mouse embryos. Whether such findings have any relevance to human intake of glycoalkaloids has yet to be determined.

There are continuing reports of human poisoning by toxic potatoes. In a detailed study of solanine poisoning among London schoolboys, McMillan & Thompson (1979) reported that potatoes containing 25 to 30 mg α-solanine/100 g peeled, boiled potato were inadvertently fed to 78 boys, who all became ill, 17 requiring hospitalization. Each boy had consumed two peeled and boiled potatoes. The authors suggested that toxic glycoalkaloids other than α-solanine could be present or another constituent might enhance alimentary absorption of solanine to toxic amounts. Steroid saponins, for instance, are known frequently to accompany *Solanum* alkaloids and could act as enhancing agents, since saponins are known emulsifiers and have been used to promote gastrointestinal absorption. McMillan & Thompson noted that the toxicity of potatoes may be due to a combination of solanidine alkaloid and spirostane saponins, although the concentrations of saponins in toxic potatoes are not known. They further speculate that the saponins could also potentiate teratogenicity.

Jadhav & Salunkhe (1975) have pointed out that most toxicity studies have focused on α-solanine, which is poorly absorbed from the gastrointestinal tract and tends to concentrate in the spleen, kidney, lung, fat, heart, brain and blood. They also note the inhibitory effects of α-solanine on cholinesterase, but emphasize that information on the pharmacology and toxicity of α-chaconine, although it represents nearly 60% of total potato alkaloids, is meagre.

One of the few studies on α-chaconine was conducted by Alozie *et al.* (1979). They report that the absorption characteristics of α-chaconine within the gastrointestinal tract are different from those of solanine, and, unlike solanine, α-chaconine is eliminated in the urine and faeces to only a small extent. ^{3}H-labelled α-chaconine in food was well absorbed from the gastrointestinal tract of hamsters, with 25% of the label excreted in seven days. Tissue concentrations of radioactivity peaked at 12 h, following oral administration, being highest in lungs, liver, spleen, skeletal muscle, kidney and pancreas; heart and brain contained only moderate amounts. In brain, liver and heart, most of the radioactivity was in the nuclear and microsomal fractions. Tissue binding studies revealed that all of the label in brain was in bound (non-extractable) form. On this basis, glycoalkaloids may act by binding to specific sites in the affected tissues.

From the above results, it is clear that the involvement of α-chaconine in health merits further investigation (Maga, 1980), as do the effects of other glycoalkaloids.

Control of accumulation

The basic recommendation to prevent glycoalkaloid accumulation is to develop and grow varieties of potato with inherently low or normal glycoalkaloid content. Beyond this, there are several agronomic and post-harvest techniques to help to control increases in glycoalkaloids.

Salunkhe & Wu (1979), in an extensive review, listed a number of water, chemical, wax and oil dip treatments for tubers that prevent glycoalkaloid build-up. Although chemical treatments are often effective,

180 *Glycoalkaloids, proteinase inhibitors and lectins*

they may also require expensive or laborious removal and are thus usually suitable only for large-scale processing enterprises. The authors recommend waxing the tubers as an effective control, especially since the coated wax is easily removed by peeling. The use of oils is an effective and inexpensive method to prevent glycoalkaloid build-up, but some oils or fats give tubers an oily appearance and many oils may become rancid over time. Immersion in water is another effective control, as is the use of detergent solutions or commercial surfactants. Such treatments create anoxia or inhibit light-induced glycoalkaloid formation.

Most of the controls described by Salunkhe & Wu (1979) are more

Figure 5.2. Potatoes sold in clear plastic bags and exposed to direct sunlight in a market in the Philippines may have increased glycoalkaloid levels.

appropriate for the processing industry and have not been examined for potential use by the small farmer, retail market and home consumer. Wood & Young (1974) propose some practical procedures appropriate to cooler climates to control glycoalkaloid levels. These recommendations could be extended for use in warmer climates as follows:

(a) Keep tubers well covered with soil during the growing period.
(b) Allow the tubers to mature before harvesting.
(c) If possible, avoid harvesting on clear, sunny days (especially in cold climates, when the temperature is near freezing, or may drop to near freezing). If harvesting cannot be avoided during hours of sunshine, remove tubers as quickly as possible from exposure to light in the field (see Figure 5.1).
(d) Do not leave potatoes to dry in the field on clear, sunny days.
(e) Discard sunburned tubers.
(f) Avoid handling methods that cause bruising or skinning of tubers.
(g) Store potatoes in the dark and keep them as cool as possible.
(h) Expose tubers to the least possible amount of light during grading and other operations.
(i) Avoid using transparent plastic bags, especially for washed and brushed potatoes. If tubers have to be displayed in transparent plastic bags, keep the length of time they are exposed to light at a minimum (Figure 5.2).
(j) Store potatoes in the market or the home in shady places away from direct sunlight (see Figure 5.1).

Proteinase inhibitors
Chemical structure and functions

Potatoes are among several major crops which contain high concentrations of proteinase inhibitors. These are proteins that inhibit or prevent the activities of the major animal pancreatic digestive proteinases including trypsin, chymotrypsin, elastase and carboxypeptidases A and B (Ryan & Hass, 1981). Potatoes are almost unique in that they contain carboxypeptidase inhibitor, previously isolated only in tomatoes and round-worms (Hass et al., 1979). In addition, they are a rich source of plasma kallikrein inhibitors (Richardson, 1977). Enterokinase-inhibiting activity was also found in tubers of two potato cultivars (Lau et al., 1980). Since enterokinase initiates the cascade reaction which activates the digestive proteinases in animals, it is expected that its inhibitor would also be a potent inhibitor of digestive proteolysis.

Ryan & Hass (1981) suggest that the potato proteinase inhibitors are similar in amino acid sequence and have evolved from a common ancestral protein. Also, by gene duplication, proteinase inhibitors may have generated new inhibiting capabilities, without losing their original inhibitor functions.

Potato tubers are one of the richest sources of proteinase inhibitors (Santarius & Belitz, 1978). More than 15% of the soluble proteins in mature tubers can compromise proteinase inhibitors (Ryan & Hass, 1981; Ryan et al., 1976). At least 13 different inhibitors have been identified and about 10 have been purified and partially characterized (Liener & Kakade, 1980). These can be differentiated, by their stability in solution for 10 min at neutral pH and 80 °C, into heat-stable and heat-labile proteins (Ryan et al., 1976).

Knowledge of the function of proteinase inhibitors in plants is meagre and speculative (Liener & Kakade, 1980). Potato proteinase inhibitors are believed to function primarily in the plant and tuber defence mechanisms against pest attack (Santarius & Belitz, 1978; Liener & Kakade, 1980) and wounding (Lau et al., 1980). While suggesting that proteinase inhibitors neutralize the effects of the digestive enzymes of invading pests or microorganisms, Hass et al. (1981) reported that carboxypeptidase inhibitor has a potent effect on all digestive tract carboxypeptidases examined.

Chymotrypsin inhibitor serves as a storage protein during potato plant development (Melville & Ryan, 1972; Ryan et al., 1976). It accumulates in leaves during insect attack, suggesting that it may be part of the defences against microorganisms or pests (Green & Ryan, 1972). Peng (1975) found that both chymotrypsin and trypsin inhibitors were present in higher levels in 'La Chipper', a potato variety with field resistance to late blight, than in 'Red LaSoda', which shows no such field resistance. Higher levels were also found in late blight resistant tomato plants, prompting the suggestion that these inhibitors play a role in the defence mechanism against this fungus.

Ryan & Green (1974) have described how proteinase inhibitors function in plant defence mechanisms. They report that insect damage or mechanical wounding of the leaves causes the release of a proteinase inhibitor including factor, which quickly moves throughout the plant causing the accumulation of inhibitors that are potentially toxic to invading insects or microorganisms. They suggest that this reaction might be a primitive immune-type response. The plant 'senses' insect attack through the release of proteinase inhibitor including factor. The inhibitors then prevent enzyme activity in attackers thus causing their death. Liener

& Kakade (1980) speculate that the puzzling inhibition of mammalian enzymes by plant proteinase inhibitors may represent an interesting side effect devoid of any true physiological significance as far as the plant is concerned.

Nutritional significance

Currently, understanding of the nutritional significance of potato proteinase inhibitors is almost non-existent. The exact nutritional significance of the inhibitors in animal and human diets is difficult to assess. According to Richardson (1977) the likelihood of a particular protein inhibitor causing adverse effects in humans depends on its ability to survive the acid conditions (pH 2 to 3) in the stomach and digestion by pepsin. Moreover, potatoes are rarely consumed in a completely raw condition by human beings.

Most plant proteinase inhibitors are destroyed by heat (Liener & Kakade, 1980), and this effect is usually accompanied by an enhancement of the nutritional value of the protein. Ryan & Hass (1981) found that boiling, micro-wave heating, or baking the tuber destroyed most inhibitors, but that the carboxypeptidase inhibitor was extremely stable to all three methods of cooking. Huang *et al.* (1981) confirmed that carboxypeptidase inhibitor is the most heat-stable potato inhibitor, and found that significant chymotrypsin inhibitor activity also survived baking and boiling, although trypsin inhibitor activity was completely destroyed. The authors added that, if the carboxypeptidase inhibitor can act on human carboxypeptidase, is still active in human intestine, and can inhibit or reduce proteolysis, then its removal by genetic selection could significantly improve the nutritional quality of cooked potatoes.

Animal feeding trials have demonstrated the effects of consuming raw, partially cooked and cooked potatoes on nitrogen utilization. Whittemore *et al.* (1975), in experiments with pigs, found that chymotrypsin inhibitor activity was high in all feeds of raw potato but absent in those containing cooked potato. They suggested that the chymotrypsin inhibitor contributed to the deleterious effects on nitrogen utilization in pigs given raw potatoes.

Pearce *et al.* (1979) found that proteinase inhibitors which were heat stable when pure but rapidly destroyed in intact tubers when cooked were rich in cysteine. The growth of chicks given experimental diets supplemented with inhibitor protein fraction from uncooked potatoes was severely depressed. Autoclaving the protein fraction resulted in normal growth. The authors concluded that selection for high protein, on the basis of high inhibitor contents, could provide nutritionally available cysteine, provided the potatoes are cooked before consumption. However, it is possible that selection for high inhibitor levels, to increase nutritional quality, could also inadvertently increase the levels of heat-stable inhibitors such as the carboxypeptidase inhibitors. Since the effect of the heat-stable inhibitors on humans is largely unknown, these aspects should be studied prior to selection for higher total inhibitor levels.

Livingston *et al.* (1980) explored the effects of raw potato extracts on the digestibility of barley in diets for pigs. They found that potato chymotrypsin inhibitor was responsible for poor nitrogen utilization. Partial cooking reduced inhibition by one-third, but steaming at 100 °C for 20 min completely destroyed inhibitor activity. Digestibility of the N from the diet with raw potato was 32.8%, while that of the heat-treated diet was 89.8%. The digestibility of N in partially cooked potatoes was 48% of that of the completely cooked sample. These experiments showed that the low nutritional value of N from raw potatoes is due to anti-nutritional factors in the tuber (the inhibitors) and not to physical inaccessibility of potato protein to enzyme attack. The authors advised that any potato preparation or processing for animal feed must utilize heat, which ruptures the potato cells and denatures most of the inhibitor, thus providing a satisfactory feed for animals.

The above discussion does not include direct nutritional effects of potato proteinase inhibitors on humans. For soybeans, considerable research has shown that the action of proteinase inhibitors may not be restricted to direct inhibition of proteolysis. The picture is confused by the fact that inhibitor effects may differ according to which type of experimental animal is used. No work has been done on the effects of potato proteinase inhibitors in human nutrition.

Cooking or processing potatoes at high altitudes could reduce the destruction of inhibitors by heat, leaving them intact and functioning even after cooking. Gursky (1969) reports that in highland areas of Nunoa, Peru, the boiling point of water is 84.9 °C rather than 100 °C and tubers are often cooked in such a way that the centre remains raw. Therefore, some of the heat-labile vitamins that are normally destroyed at sea level may not be destroyed at higher altitudes. The same could be true for proteinase inhibitors.

Lectins

Lectins, or haemagglutinins, are carbohydrate-binding cell-agglutinatory proteins which occur widely in the plant and animal kingdom. Goldstein & Hayes (1978) noted that lectins have played an important role in the development of immunology and are currently used in serological laboratories for typing blood and determining secretor status, separating leucocytes from erythrocytes, and agglutinating cells from blood in the preparation of plasma. Lectins also serve as reagents for the detection, isolation and characterization of blood-group antigens. Since some lectins have been found to distinguish normal from malignant cells, the possible use of lectins in cancer chemotherapy has been suggested (Goldstein & Hayes, 1978). Bean (*Phaseolus vulgaris*) lectins have received considerable attention due to the toxic effects of these beans when undercooked (Putsztai et al., 1981).

The lectin in potato tubers was first described in 1926 and has since been the subject of several detailed studies (Kilpatrick, 1980). It agglutinates erythrocytes of several mammalian species including human (irrespective of blood-group type), guinea pig, mouse, rabbit and sheep (Goldstein & Hayes, 1978). Recently, a lectin whose characteristics resemble those of the tuber lectin has been isolated from potato fruits (Kilpatrick, 1980).

Little is known about the physiological role of these two lectins in the potato plant or indeed about the function of plant lectins in general. Kilpatrick (1980) suggests that, since the tuber is in part a storage tissue, it is possible that the lectin functions in some process concerned with the storage, maintenance or utilization of food or food reserves. Another possibility is that it has a defence function similar to that of wheat-germ agglutinin, which is thought to protect plants against chitin-containing phytopathogens. The potato tuber lectin has a saccharide specificity similar to that of wheat-germ agglutinin and might also be able to inhibit fungal growth (Kilpatrick, 1980). No physiological function has yet been postulated for the lectin discovered in potato fruits.

Virtually nothing has been written about the nutritional significance or the potential toxic effects of the potato lectins. Since the toxic effects of lectins found in various bean species have been so thoroughly documented (Goldstein & Hayes, 1978), it might be anticipated that under certain conditions, potato lectins could also be toxic. They are, however, heat labile. Further research is needed to clarify the function of the potato lectin and its nutritional significance for humans.

Summary

Although there is considerable evidence concerning the toxicity of glycoalkaloids to human beings, the toxic nature of proteinase inhibitors and that of lectins are largely inferred from studies of other foods and from feeding experiments with animals. The toxicity of some glycoalkaloids, notably α-chaconine, may be related to their ability to bind to certain tissues, and pregnant mammals are more susceptible to total glycoalkaloids.

Cooking and processing have differential effects on the toxic components of potatoes: some proteinase inhibitors are destroyed by normal cooking, but others are heat-stable. The nutritional significance of the latter is not yet known. Glycoalkaloids are not destroyed by cooking or processing but, because of their location in the tuber, are mainly removed on peeling. However, when potatoes have unusually high levels, or when wounding or damage has occurred, peeling may only remove 30%.

It has been recommended that efforts be made to breed and select new varieties with low to moderate amounts of glycoalkaloids. Additionally, precautions can be taken to avoid the conditions which provoke increases in glycoalkaloid levels. However, most evidence points to a genetically controlled role of both glycoalkaloids and proteinase inhibitors in plant and tuber defence mechanisms against pests or microorganisms. Increasing concern over growing pesticide use and greater resistance of pests to potent pesticides provide good reasons for research efforts to explore the potential manipulation of glycoalkaloids and proteinase inhibitors to enhance resistance to pests and microorganisms. However, such efforts must be balanced with controls to ensure that toxicity is not an inadvertent result.

As a final note, if potatoes are purposely subjected to hostile environments, greater pest attack or poor storage and handling, in the hopes of extending current growing seasons or creating new ones, then greater attention should be paid to the potential toxicity to be caused by the reactions of the plant's own defence mechanisms to these conditions.

References

Ahmad, R. (1982). Survey of glycoalkaloid content in potato tuber growing in Pakistan and study of environmental factors causing their synthesis, and physiological investigations on feeding high glycoalkaloid greened potatoes to experimental animals. *Sixth Annual Research Report*. Department of Botany, University of Karachi, Pakistan.

Ahmed, S. S. & Müller, K. (1978). Effect of wound-damage on the glycoalkaloid content in potato tubers and chips. *Lebensm.-Wiss. Technol.* **11**: 144–6.

Allen, E. H. & Kuć, J. (1968). α-Solanine and α-chaconine as fungitoxic compounds in extracts of Irish potato tubers. *Phytopathology* **58**: 776–81.

Alozie, S. O., Sharma, R. P. & Salunkhe, D. K. (1979). Physiological disposition, subcellular distribution and tissue binding of α-chaconine (^{3}H). *J. Food Safety* **1**, 257–273.

Anon. (1975). End of the potato avoidance hypothesis. *Br. Med. J.* **4**: 308–309.

Antúnez de Mayolo, S. E. (1981). [*Nutrition in ancient Peru.*] In Spanish. Banco Central de Reserva del Perú, Lima.

References

Arutyunyan, L. A. (1940). The solanine content of potatoes. *Vop. Pitan.* **9**: 30–6.
Baerug, R. (1962). Influence of different rates and intensities of light on solanine content and cooking quality of potato tubers. *Eur. Potato J.* **5**: 242–51.
Baker, L. C., Lampitt, L. H. & Meredith, O. B. (1955). Solanine glycoside of the potato. III. An improved method of extraction and determination. *J. Sci. Food Agric.* **6**: 197–202.
Bergers, W. W. A. (1980). A rapid quantitative assay for solanidine glycoalkaloids in potatoes and industrial potato protein. *Potato Res.* **23**: 105–10.
Bushway, R. J., Bureau, J. L. & McGann, D. F. (1983). Alpha-chaconine and alpha-solanine content of potato peels and potato peel products. *J. Food Sci.* **48**: 84–6.
Bushway, R. J., Bushway, A. A. & Wilson, A. M. (1981). α-Chaconine and α-solanine content of MH:30 treated Russet Burbank, Katahdin and Kennebec tubers stored for nine months at three different temperatures. *Am. Potato J.* **58**: 498.
Bushway, R. J. & Ponnampalam, R. (1981). α-Chaconine and α-solanine content of potato products and their stability during several modes of cooking. *J. Agric. Food Chem.* **29**: 814–17.
Christiansen, J. A. (1977). The utilization of bitter potatoes to improve food production in the high altitude of the tropics. Ph.D. thesis. University of Cornell, Ithaca, NY.
Christiansen, J. A. & Thompson, N. R. (1976). [The utilization of 'bitter' potatoes in the cold tropics of Latin America.] In Spanish. *Proceedings of the 4th Symposium of the International Society for Tropical Root Crops, Cali, Colombia.*
Coxon, D. T. & Jones, P. G. (1981). A rapid screening method for the estimation of total glycoalkaloids in potato tubers. *J. Sci. Food Agric.* **32**: 366–70.
Coxon, D. T., Price, K. R. & Jones, P. G. (1979). A simplified method for the determination of total glycoalkaloids in potato tubers. *J. Sci. Food Agric.* **30**: 1043–9.
Deahl, K. L., Young, R. J. & Sinden, S. L. (1973). A study of the relationship of late blight resistance to glycoalkaloid content in fifteen potato clones. *Am. Potato J.* **50**: 248–53.
Fitzpatrick, T. J., Herb, S. F., Osman, S. F. & McDermott, J. A. (1977). Potato glycoalkaloids: increases and varieties of ratios in aged slices over prolonged storage. *Am. Potato J.* **54**: 539–44.
Fitzpatrick, T. J., McDermott, J. A. & Osman, S. F. (1978). Evaluation of injured commercial potato samples for total glycoalkaloid content. *J. Food Sci.* **43**: 1617–18.
Goldstein, I. J. & Hayes, C. E. (1978). The lectins: carbohydrate-binding proteins of plants and animals. In R. S. Tipson & D. Horton (eds.), *Advances in carbohydrate chemistry and biochemistry*, vol. 35. Academic Press, New York.
Green, T. & Ryan, C. A. (1972). Wound-induced proteinase inhibitor in plant leaves: a possible defense mechanism. *Science* **175**: 776–7.
Gregory, P., Sinden, S. L., Osman, S. F., Tingey, W. M. & Chessin, D. A. (1981). Glycoalkaloids of wild, tuber-bearing *Solanum* species. *J. Agric. Food Chem.* **29**: 1212–15.
Gull, D. D. & Isenberg, F. M. (1960). Chlorophyll and solanine content and distribution in four varieties of potato tubers. *Am. Soc. Hort. Sci.* **75**: 545–56.
Hass, G. M., Ager, S. P., Le Tourneau, D., Derr-Makus, J. E. & Makus, D. J. (1981). Specificity of the carboxypeptidase inhibitor from potatoes. *Plant Physiol.* **67**: 754–8.

Hass, G. M., Derr, J. E., Makus, D. J. & Ryan, C. A. (1979). Purification and characterization of the carboxypeptidase isoinhibitor from potatoes. *Plant Physiol.* **64**: 1022–8.

Henfling, J. W. (1979). Aspects of the elicitation and accumulation of terpene phytoalexins in the potato – *Phytophthora infestans* interaction. Ph.D. thesis. University of Kentucky.

Huang, D. Y., Swanson, B. G. & Ryan, C. A. (1981). Stability of proteinase inhibitors in potato tubers during cooking. *J. Food Sci.* **46**: 287–90.

Jadhav, S. J. & Salunkhe, D. K. (1975). Formation and control of chlorophyll and glycoalkaloids in tubers of *Solanum tuberosum* L. and evaluation of glycoalkaloid toxicity. *Adv. Food Res.* **21**: 307–54.

Kilpatrick, D. C. (1980). Isolation of a lectin from the pericarp of potato (*Solanum tuberosum*) fruits. *Biochem. J.* **191**: 273–5.

Kuć, J. (1972). Phytoalexins. *Annu. Rev. Phytopathol.* **10**: 204–32.

Lau, A., Ako, H. & Werner-Washburne, M. (1980). Survey of plants for enterokinase inhibitors. *Biochem. Biophys. Res. Commun.* **92**: 1243–9.

Liener, I. E. & Kakade, M. L. (1980). Protease inhibitors. In I. E. Liener (ed.), *Toxic constituents of plant foodstuffs*, 2nd edn. Academic Press, New York.

Linnemann, A. R., van Es, A. & Hartmans, K. J. (1985). Changes in the content of L-ascorbic acid, glucose, fructose, sucrose and total glycoalkaloids in potatoes (cv. Bintje) stored at 7, 16 and 28 °C. *Potato Res.* **28**: 271–8.

Livingstone, R. M., Baird, B. A., Atkinson, T. & Crofts, R. M. J. (1980). The effect of either raw or boiled liquid extract from potato (*Solanum tuberosum*) on the digestibility of a diet based on barley in pigs. *J. Sci. Food Agric.* **31**: 695–700.

Maga, J. A. (1980). Potato glycoalkaloids. *CRC Crit. Rev. Food Sci. Nutr.* **12**: 371–405.

Maga, J. A. (1981). Total and individual glycoalkaloid composition of stored potato slices. *J. Food Processing Preserv.* **5**: 23–9.

McCollum, G. D. & Sinden, S. L. (1979). Inheritance study of tuber glycoalkaloids in a wild potato. *Solanum chacoense* Bitter. *Am. Potato J.* **56**: 95–113.

McMillan, M. & Thompson, J. C. (1979). An outbreak of suspected solanine poisoning in schoolboys: examination of criteria of solanine poisoning. *Q. J. Med., New Series*, **48**: 227–43.

Melville, J. C. & Ryan, C. A. (1972). Chymotrypsin inhibitor I from potatoes: large scale preparation and characterization of its subunit components. *J. Biol. Chem.* **247**: 3445–53.

Morgan, M. R. A., McNerney, R., Matthew, J. A., Coxon, D. T. & Chan, H. W. S. (1983). An enzyme-linked immunosorbent assay for total glycoalkaloids in potato tubers. *J. Sci. Food Agric.* **34**: 593–8.

Morgan, M. R. A., Coxon, D. T., Bramham, S., Chan, H. W.-S., van Gelder, W. M. J. & Allison, M. J. (1985). Determination of the glycoalkaloid content of potato tubers by three methods including enzyme-linked immunosorbent assay. *J. Sci. Food Agric.* **36**: 282–8.

Morris, S. C. & Lee, T. H. (1981). Analysis of potato glycoalkaloids with radically compressed high-performance liquid chromatographic cartridges and ethanolamine in the mobile phase. *J. Chromatogr.* **219**: 403–10.

Morris, S. C. & Lee, T. H. (1984). The toxicity and teratogenicity of Solanaceae glycoalkaloids, particularly those of the potato (*Solanum tuberosum*): a review. *Food Technol. Australia* **36**: 118–24.

Nair, P. M., Rehere, A. G. & Ramaswamy, N. K. (1981). Glycoalkaloids of *Solanum tuberosum* Linn. *J. Sci. Ind. Res.* **40**: 529–35.

Osman, S. F., Herb, S. F., Fitzpatrick, T. J. & Schmiediche, P. (1978).

Glycoalkaloid composition of wild and cultivated tuber-bearing *Solanum* species of potential value in potato breeding programs. *J. Agric. Food Chem.* **26**: 1246–8.

Osman, S. F., Herb, S. F., Fitzpatrick, T. J. & Sinden, S. L. (1976). Commersonine, a new glycoalkaloid from two *Solanum* species. *Phytochemistry* **15**: 1065.

Osman, S. F. & Sinden, L. (1977). Analysis of mixtures of solanidine and demissidine glycoalkaloids containing identical carbohydrate units. *J. Agric. Food Chem.* **25**: 955–7.

Pearce, G., McGinnis, J. & Ryan, C. A. (1979). Utilization by chicks of half-cystine from native and denatured proteinase inhibitor protein from potatoes. *Proc. Soc. Exp. Biol. Med.* **160**: 180–4.

Peng, J. H. (1975). Increased activity of proteinase inhibitors in disease resistance to *Phytophthora infestans*. *Diss. Abstr. Int. B* **36**: 24–B.

Pierro, L. J., Haines, J. S. & Osman, S. F. (1977). Teratogenicity and toxicity of purified alpha chaconine and alpha solanine. *Teratology* **15**: 31 A.

Ponnampalam, R. & Mondy, N. I. (1983). Effect of cooking on the total glycoalkaloid content of potatoes. *J. Agric. Food Chem.* **31**: 493–5.

Pusztai, A., Clarke, E. M. W., Grant, G. & King, T. P. (1981). The toxicity of *Phaseolus vulgaris* lectins. Nitrogen balance and immunochemical studies. *J. Sci. Food Agric.* **32**: 1037–46.

Raman, K. V., Tingey, W. M. & Gregory, P. (1979). Potato glycoalkaloids: effects on survival and feeding behavior of the potato leafhopper. *J. Econ. Entomol.* **72**: 337–41.

Renwick, J. H. (1972). Anencephaly and spina bifida are usually preventable by avoidance of a specific but unidentifiable substance present in certain potato tubers. *Br. J. Prev. Soc. Med.* **26**: 67–88.

Renwick, J. H. (1973). Prevention of anencephaly and spina bifida in man. *Teratology* **8**: 321–33.

Rhoades, R. E. (1982). The incredible potato. *Natl. Geogr. Mag.* **161**: 668–94.

Richardson, M. (1977). The proteinase inhibitors of plants and micro-organisms. *Phytochemistry* **16**: 159–69.

Robinson, T. (1974). Metabolism and function of alkaloids in plants. *Science* **184**: 430–5.

Ross, H., Pasemann, P. & Nitzsche, W. (1978). [Glycoalkaloid content of potatoes and its relationship to location, year and taste.] In German. *Z. Pflanzenzuecht.* **80**: 64–79.

Ryan, C. A. & Green, T. R. (1974). *Proteinase inhibitors in natural plant protection*, Scientific paper no. 4122. College of Agric. Res. Center, Washington State University.

Ryan, C. A. & Hass, G. M. (1981). Structural, evolutionary and nutritional properties of proteinase inhibitors from potatoes. In R. L. Ory (ed.), *Antinutrients and natural toxicants in foods*. Food and Nutrition Press Inc., Westport, CT.

Ryan, C. A., Kuo, T., Pearce, G. & Kunkel, R. (1976). Variability in the concentration of 3 heat stable proteinase inhibitor proteins in potato tubers. *Am. Potato J.* **53**: 443–5.

Salunkhe, D. K. & Wu, M. T. (1979). Control of post harvest glycoalkaloid formation in potato tubers. *J. Food Protection* **42**: 519–25.

Salunkhe, D. K., Wu, M. T. & Jadhav, S. J. (1972). Effects of light and temperature on the formation of solanine in potato slices. *J. Food Sci.* **37**: 969–70.

Sanford, L. L. & Sinden, S. L. (1972). Inheritance of potato glycoalkaloids. *Am. Potato J.* **49**: 209–17.
Santarius, K. & Belitz, H. D. (1978). Proteinase activity in potato plants. *Planta* **141**: 145–53.
Schwarze, P. (1962). [Methods for identification and determination of solanine in potato breeding material.] In German. *Züchter* **32**: 155–60.
Shaw, R. & Booth, R. (1982). *Simple processing of dehydrated potatoes and potato starch.* International Potato Center, Lima.
Sikilinda, V. A. & Kiryukhin, V. P. (1975). [Investigation of glycoalkaloids of potatoes.] In Russian. *Nauchnye Trudy, Institut Kartofel'nogo Khozyastva*, no. 21: 5–11.
Sinden, S. L., Deahl, K. L. & Aulenback, B. B. (1976). Effect of glycoalkaloids and phenolics on potato flavor. *J. Food Sci.* **41**: 520–3.
Sinden, S. L., Goth, R. W. & O'Brian, M. J. (1973). Effect of potato alkaloids on the growth of *Alternaria solani* and their possible role as resistance factors in potatoes. *Phytopathology* **63**: 303–7.
Sinden, S. L. & Webb, R. W. (1974). *Effect of environment on glycoalkaloid content of six potato varieties at 39 locations*, Technical Bulletin no. 1472. US Dept. of Agriculture, Agric. Res. Serv.
Sizer, C. E., Maga, J. A. & Craven, C. J. (1980). Total glycoalkaloids in potatoes and potato chips. *J. Agric. Food Chem.* **28**: 578–9.
Smittle, D. A. (1971). A comparison and modification of methods of total glycoalkaloid analysis. *Am. Potato J.* **48**: 410–13.
Thomas, P. (1984). Radiation preservation of foods of plant origin. Part 1. Potatoes and other tuber crops. *CRC Crit. Rev. Food Sci. Nutr.* **19**: 327–79.
Tingey, W., McKenzie, J. D. & Gregory, P. (1978). Total foliar glycoalkaloids and resistance of wild potato species to *Empoasca fabae* (Harris). *Am. Potato J.* **55**: 577–85.
Werge, R. W. (1979). Potato processing in the central highlands of Peru. *Ecol. Food Nutr.* **7**: 229–34.
Whittemore, C. T., Taylor, A. G., Moffat, I. W. & Scott, A. (1975). Nutritional value of raw potato for pigs. *J. Sci. Food Agric.* **26**: 255–60.
Wolf, M. J. & Duggar, B. M. (1946). Estimation and physiological role of solanine in the potato. *J. Agric. Res.* **73**: 1–32.
Wood, F. A. & Young, D. A. (1974). *TGA in potatoes*, Canada Department of Agriculture Publication no. 1533.
Wu, M. T. & Salunkhe, D. K. (1976). Changes in glycoalkaloid content following mechanical injuries to potato tubers. *J. Am. Soc. Hort. Sci.* **101**: 329–31.
Zaletskaya, B. G., Golynskaya, L. A. & Zaletskii, V. N. (1977). [Changes in the contents of glycoalkaloids during the manufacture and storage of dried potato purée.] In Russian. *Konservn. Ovoshchesush. Promst.* no. 1: 13–14.
Zitnak, A. (1964). The significance of glycoalkaloids in the potato plant. *Proc. Can. Soc. Hort. Sci.* **3**: 81–6.
Zitnak, A. & Johnston, G. R. (1970). Glycoalkaloid content of B5141-6 potatoes. *Am. Potato J.* **47**: 256–60.

6

Patterns of potato consumption in the tropics

The potato has spread around the world during the past 400 years and adapted to a wide variety of environments and an equally diverse range of human tastes and preferences. In some tropical developing countries, it is a common vegetable, while elsewhere, consumption ranges from 1 kg to more than 100 kg per person annually. To some, the potato is the 'bread of life', while to others it is taboo. This chapter addresses the great diversity in potato consumption patterns. It discusses important issues concerning the potato's role in developing countries and its potential for the future.

There is a growing realization amongst food planners that 'programmes aimed at increasing the production of food, even if they are successful, must be accompanied by efforts designed to affect the distribution of incomes and patterns of diet' (Berg, 1981). The potential for increasing consumption of a food item is largely determined by the extent to which its role in the diet can be altered according to changes in supply or cost. Hence, it is essential to consider not only the production, storage and marketing of the potato, but also consumption behaviour.

This chapter was written with the following questions in mind:

1. How much potato is currently consumed in the tropics and by whom?
2. How are potatoes consumed and what factors regulate potato preferences and consumption patterns?
3. What is the potential for greater potato utilization?

It is hoped that the answers suggested below will stimulate other researchers to probe more deeply into the complexities governing potato consumption in developing countries.

Calculating consumption
Food balance sheets

The most frequently used method of calculating potato consumption is the food balance sheet (FBS). Such sheets show total supply of food available in a country for a given year (or period of years) against the itemized utilization of those foodstuffs (Farnsworth, 1961). This can be summarized as:

Potatoes available for human consumption = Domestic production, imports, exports, and net changes in year-end stocks − Quantities used for seed, non-food industry, animal feed and waste

The equation is usually developed to show the calories and grams of protein provided by potatoes per capita per day. Unfortunately, 'per capita *availability*' is often referred to as 'per capita *consumption*'. This is erroneous for the following reasons. The information fed into the equation is often unreliable. Determining total production of potatoes is difficult: because the plants are usually grown in more isolated areas, in highland and mountainous regions, and in small discontinuous plots, official measurements of production areas are inaccurate. Potatoes are often grown more than once a year, yet are counted only once. They may be grown as intercrops, multicrops, relay crops, secondary crops, or backyard garden crops, and as such, are overlooked in national tallies. In northern Luzon, in the Philippines, for example, yields were officially estimated at 6 to 8 tons/ha (1 ton = 1.016 tonnes). But when yield surveys of farmers were carried out, it was found that the actual average is 28 tons/ha (Potts, 1981).

Utilization data are also suspect. In many cases, seed production rates in the tropics are estimated on the basis of temperate zone rates, but due to the widespread use of marble-sized seed in many countries, seed tonnage may be overestimated. What is considered to be potato waste differs widely and standard figures cannot be applied to all developing countries. Waste factors of 10%, 15% or even 20% are applied indiscriminately without regard to local habits of peeling, cooking or peel utilization. Many populations consume large quantities of peel, leaving little, if any, waste. Waste determination is further complicated when peels or rotten potatoes are fed to animals that are subsequently consumed (Figure 6.1). In highland Peru, small shrivelled potatoes, 'waste' by North American or European standards, are consumed and prized for their sweeter taste' (Werge, 1979). Poor-quality potatoes, classed as

Calculating consumption

waste in developed country markets, are sold at low prices in the developing countries. 'Waste' potatoes may be 'donated' to local institutions, such as orphanages, or 'collected' by local beggars. Partially rotten potatoes are cut, peeled and sold in plastic bags, often at higher prices per kilogram than whole potatoes.

Total national population size can also be underestimated, or overestimated, in developing countries.

Figure 6.1. Bhutanese potato producers feed their pigs potatoes that are not suitable for family consumption, thus converting a 'waste' product to one of higher value.

In light of the dubious quality of the data, it seems doubtful that the resulting per capita availability, is of any value. However, even if all the inputs are accurate, the result cannot be called 'per capita consumption'. The latter term implies equal food distribution among all ethnic, religious, socio-economic, age and sex groups. This does not happen. Additionally, FBSs cannot account for seasonal or regional differences in consumption among these groups. In sum, if only the FBS is used, it is difficult to calculate current consumption levels adequately, to estimate the potential for change in consumption rates or to identify target groups within a country for specific consumption-related programmes.

Nutrition surveys

These surveys are conducted at national, regional and local levels. They are useful for determining potato consumption on a daily basis, and, depending on the quality of the survey, to distinguish patterns of consumption according to regional classifications, socio-economic status, price fluctuations, or seasonal availability. Total consumption of a large sample of people for a period of time (24 h, three days, or a week) is used to calculate daily intake averages and annual consumption rates for the area under study. If surveys are conducted with the same sample population at different times during a year, seasonal consumption of certain items may be ascertained.

Although nutrition surveys supply a wealth of information on potato consumption, there are three problems which limit their usefulness. (1) Their goal is to measure caloric and protein levels in the subject population to determine dietary adequacy. Often foods are grouped together and their nutritional contribution is considered as a unit – for example, root and tuber crops. Depending upon the level where the grouping took place, it may or may not be possible to separate potato consumption from that of other foods. (2) If the informant has to recall a meal, and potato played only a nominal role in total intake, any potatoes consumed may be forgotten. This is especially true if the recall period is longer than 24 h. (3) These methods cannot measure adequately potatoes consumed only on special or ritual occasions, since nutrition surveys normally take place on working days and not on Sundays, holidays, holy days or festival days. It is true that consumption of luxury or special foods may not have high nutritional significance, but estimating their periodic demand so as to plan for production, marketing and storage is important.

Both FBSs and nutrition surveys should be used as the starting point for calculating potato consumption. Despite their limitations, they can provide indications of availability and some consumption trends, and they form the basis of most international and national statistics. To make correct decisions on priority areas for potato promotion, therefore, more reliable estimates of consumption are needed, as well as an understanding of why people do or do not eat potatoes. Consumer responses to potato price, size, colour, quality and origin and inclinations towards altering their potato consumption must be determined.

Consumer surveys and consumer groups

Numerous methods and manuals are used for conducting consumer surveys (Den Hartog & van Staveren, 1979; Christakis, 1978; Chavez & Huenemann, 1981). Time, budget, and personnel constraints usually set limits on the depth and extent of any survey, but generally enough information, at least to make initial planning decisions, can be gathered with informal surveys (Rhoades, 1982; Ruano & Calderón, 1981). These take the form of open, spontaneous dialogues and conversations with a variety of informants (Figure 6.2); they eliminate costly data processing and the normal timelag between survey and results and recommendations.

Whatever interview method is employed, the most useful procedure is to distinguish the types of consumer along the chain connecting production zones, markets, and urban and rural households. Examination of each link in the chain allows identification of homogeneous consumer groups, their average rate and frequency of consumption and the constraints or factors which regulate their consumption. Specific consumer types vary from country to country but as a minimum, the national population should be grouped according to agroecological zones and urban or rural residence. Some examples are described below.

Potato producers

Farm size, land tenancy, and the number of potato crops planted per year will distinguish sub-divisions within this group. Consumption can vary a great deal among potato producers. Some store large quantities to last all year until the following harvest. Others store only small quantities and produce a crop several times a year, supplementing their own stocks with supplies from the market-place. Some do not store at all, consuming potatoes only at harvest time. Though some potato farmers are said to be strictly commercial producers, it is rare to find a producer who is not a consumer as well.

196   *Patterns of potato consumption*

Figure 6.2. Learning about potato consumption by talking to a consumer family in their household compound in Joydebpur, Bangladesh (top), and a market vendor and clients in Bukittinggi market, West Sumatra, Indonesia (bottom).

Non-potato farmers in potato production zones

This group has greater access to potatoes than do rural non-producers living in non-production areas, since they can harvest potatoes and receive a portion of their payment in kind, exchange their production for potatoes, or purchase directly from neighbours at low prices.

Non-potato farmers outside potato production zones

These people theoretically have the least access to potatoes, must pay the highest prices and perhaps have the least familiarity with potato consumption, especially if they live at great distances from the influences of urban dietary habits. It is necessary to determine the relative importance of price and availability or social factors in limiting their consumption.

Potato marketers

Other than potato producers, potato wholesalers, retailers and small vendors have the greatest and most regular access to potatoes at the lowest cost. This group is probably the best indicator for the consumption in a non-producing population, if constraints of price and availability are removed. Additionally, sellers are often the best informants on the buying habits and preferences of the larger study populations.

Urban residents in potato production zones

Potato prices and seasonal availability patterns can differ here from urban areas in non-producer zones. Market centres in production zones usually have lower prices than in other cities, especially at harvest times, and the residents use potato in the diet. The consumption levels of this group could also indicate the overall potential urban consumption if prices are acceptable.

Urban residents in non-potato production zones

People living in small urban settings normally receive smaller proportions of the potatoes marketed than do those in regional or national capitals. Dwellers in capital cities may also be more motivated to eat potatoes as a result of the influence of western expatriate residents (see Special consumer groups, below).

Stratification of the above groups by socio-economic status will reveal further differences in the potato consumption pattern and provide a rough estimate of the amount of income which can be directed to buying potatoes for food. Generally, in the developing countries, excluding Andean South America, potato consumption is higher among the wealthier people of a particular consumer group. However, it is not known if this reflects a specific preference for more potatoes, as income levels increase, or if it indicates general increases, with higher incomes, in all foodstuffs.

Special consumer groups
In every country there are special groups which, because of economic, social, cultural or political factors, demonstrate potato consumption patterns that differ greatly from those of the groups described above.

Expatriates
Certain expatriates living in the tropical developing countries have a set of food habits which are accommodated to the new environment, but nevertheless, are not radically altered from their form in the country of origin. Potato eaters removed from areas of potato production will still maintain high levels of potato consumption, even when prices are high. Often potato producers, merchants and agricultural planners will claim (without justification) that expatriates are their major clients. The extent of their influence, past and present, on native food habits, especially in colonial situations, should be evaluated separately.

Ethnic groups
Potato consumption may vary in groups sharing the same agroecological zone, urban residence and socio-economic classification but differing in ethnic origin. In Indonesia, the ethnic Chinese are found to eat more potato than non-Chinese city residents of the same socio-economic level. Another ethnic group, the Minangkabau people of Western Sumatra, consume large quantities of marble-sized potatoes in preparations not common to the rest of Indonesia.

Pre-school children
Potatoes are used as a weaning food or a supplement to breast feeding in, for example, Guatemala, Rwanda and the Philippines. In some cases, they are fed exclusively to young children, while the rest of the family eats other foods. Often such consumption is excluded or overlooked in household surveys. The ages children are fed potatoes, types of preparation, taboos, and frequency of consumption should all be included in the survey.

Hotels and restaurants
These can constitute the major bulk buyers of potatoes. They must meet the preferences of special consumers, such as tourists, which may differ

from those of national consumers. However, care must be taken not to focus potato improvement plans only on the hotel and restaurant trade.
Nutritional and agricultural professionals
Taken separately, this group has the greatest access to accurate information about the use of the potato. Their consumption habits reveal the effects of education and exposure on the patterns of potato consumption.

Not every country will have all of the groups described above, nor indeed will the variations between official per capita availability and consumer survey results always be of the same magnitude, as the examples in the following section illustrate.

Results of consumer surveys in selected countries
Indonesia

FBSs constructed in 1978 for Indonesia, the fifth highest population in the world (149 million), reported a per capita availability of 1.35 kg/potato per year (Indonesia, 1978). A nation-wide household survey (Susenas, 1980) reported average consumption rates that were lower (see Table 6.1) for the country as a whole but higher for urban than for rural areas. Separating the household survey results for only one of Indonesia's islands, Java (with 60% of the national population), rural and urban consumptions were reported to be lower than in Indonesia as a whole.

Because the national statistics are so low, it is widely believed that potatoes play an unimportant role in the national diet and economy. Yet

Table 6.1. *Potato consumption in Indonesia (1978)*

	kg/person per year
Food balance sheets	1.35
National household survey	
Average	1.04
Rural	1.04
Urban	2.08
Java household survey	
Average	1.04
Rural	0.52
Urban	1.56

Patterns of potato consumption

>40% of the research budget of the vegetable research section of the Ministry of Agriculture for 1981–82 was allocated to potatoes. Better information on potato utilization was needed to determine if this budget expenditure was justified.

In 1981, informal consumer surveys in West Java province were conducted to identify homogeneous consumer groups and to compare their rates of potato consumption with the reported statistics. Surveys were in a highland zone producing vegetables and potatoes, a lowland rice-growing zone, and several urban centres including the national capital, Djakarta (see Figure 6.3). Some of the survey results are presented in Figure 6.4.

None of the groups identified had a per capita consumption level equal to the estimated per capita availability. In both urban and rural areas, the wealthier residents consumed greater amounts of potato than did poorer people. Interestingly, poorer lowland farmers consumed a quantity similar to that of poorer residents of urban areas or of potato production zones, indicating that lower prices and greater availability may not automatically encourage greater potato consumption. Interviews with expatriates showed that their average consumption levels are higher than those of any other group, but even so, they are lower by half than the average consumption levels in their home countries.

Potato farmers consume the highest levels; again, wealthier farm families consume more than the poorer ones. In some cases this is due to

Figure 6.3. Potato consumption study sites in West Java.

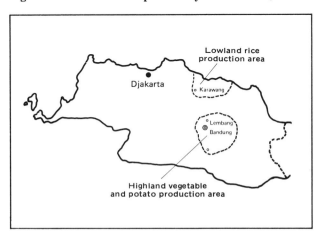

Results of consumer surveys

Figure 6.4. Consumption levels of different groups in West Java.

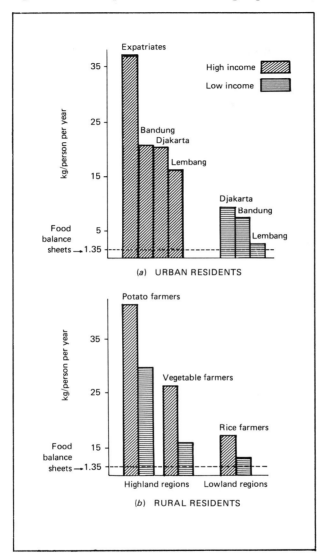

the latter's lack of sufficient storage facilities, but for many the reason is because they sell the larger portion of the crop to provide family income. Wealthier farmers also sell the majority of their crop, but they can afford to purchase potatoes when home supplies run out.

This kind of information identifies ware storage needs and the groups who might benefit most from initiating potato cultivation. It also supports the need to examine production statistics, since, as consumption levels are much higher than was previously estimated, production must have been under-reported and should be re-examined.

Rwanda

Situated in the African highlands in the watershed of the Nile and Zaïre rivers, the tiny country of Rwanda (five million inhabitants) reports the highest per capita potato availability in Africa, 46 kg/year (International Potato Center, 1982). Results of consumer surveys in 1980 throughout the country show that in production zones, areas contiguous to production zones, and in urban centres, average consumption is a great

Figure 6.5. Major agroecological zones of Rwanda.

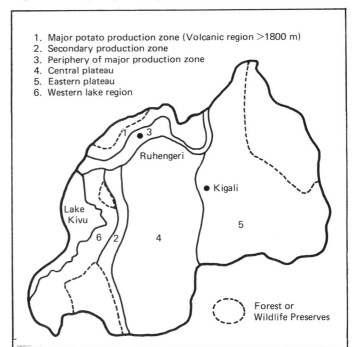

deal higher than 46 kg/year, while for other regions consumption is quite low (Figure 6.5; Table 6.2). The average for two major urban centres was also high at 153 kg/person per year.

As in Indonesia, this information indicated higher production than currently reported, and corroborated field observations of higher than average yields as well as extensive use of potatoes as an intercrop with non-food cash crops such as pyrethrum. Although only about 10% of the national population live in urban centres, the percentage is growing and more potatoes will be needed to supply increasing demands. Potato programme leaders previously thought that the inhabitants of the central and eastern plateau regions did not consume potatoes, yet consumption, although comparatively low, does occur. Informants in these areas reported a desire to consume more potatoes yet they had disappointing results in attempts to produce their own crops. Shifting some priority in the national potato programme to the introduction of new production techniques in such areas would have considerable potential impact on increasing household consumption levels.

Guatemala

Guatemala, with a population of 6.6 million, is the largest potato producer in Central America. Though potatoes have been produced in the Guatemalan highlands for centuries, per capita availability is reportedly quite low, only 4.5 kg/year (Christiansen, 1980). Nationwide nutrition surveys (INCAP/ICNND, 1971) indicate that consumption is probably much higher (Table 6.3).

Table 6.2. *Potato consumption in Rwanda*

Zone[a]	kg/person per year
1	261
2	125
3	72
4	28
5	26
6	NA

NA, not available.
[a] Numbers 1 to 6 correspond to the major agroecological zones shown in Fig. 6.5.

Table 6.3. *Potato consumption in Guatemala, 1965–67*

Location	kg/person per year
Urban	13.32
Rural	6.0
Average	8.32

Source: INCAP/ICNND (1971).

Patterns of potato consumption

Table 6.4. *Guatemala City: urban potato consumption (1981)*

Income group	kg/person per year
High	36.9
Middle	35.8
Low	25.1
Average	28.7

Looking only at urban consumption in the capital, Guatemala City, consumer surveys in 1981 (Table 6.4) showed much higher consumption levels among each of the three income groups examined than Christiansen's (1980) estimate and that of INCAP/ICNND (1971). It is unlikely that potato consumption has vastly increased, but focusing on potatoes exclusively in a survey provides more accurate consumption recall than when all elements of the diet must be recollected by the informant. There was little difference between consumption levels of middle- and of high-income groups, but the low-income group consumed 10 kg/person per year less. The majority of the informants in the last category expressed a desire to consume more potatoes, but said that cost prohibited them from doing so. Most of the upper-income informants, stated that what they currently consumed was sufficient. This indicates a potential among low-income groups to consume more if prices could be lowered, but that this potential has a ceiling or level of sufficiency at around 35 kg/person per year.

Peru

Per capita availability was estimated at 63 kg in Peru for 1981 (Scott, 1981). Despite the fact that this country is well-known for its high levels of potato consumption, there is a great deal of variation within the population, which can be seen if the four major agroecological zones are examined individually (Figure 6.6).

In the Peruvian Andean highlands, potatoes are a primary food source and are consumed on a daily basis of 0.5 kg/person per day or more. Annual consumption per capita of 100 to 200 kg is common and a wide range of traditional varieties is produced to meet home demands for different preparations and tastes.

Results of consumer surveys

In the eastern mid-elevation zone, potatoes are treated as vegetables, complementing other locally produced roots and tubers, as well as a variety of fruits and vegetables. Although potatoes are not produced extensively in this zone, they are readily available in markets, or can be obtained, by trading, from neighbouring highland zones. Consumption levels vary from 10 to 100 kg/person per year, depending on the frequency of family access to a potato source.

Figure 6.6. Agroecological zones of Peru.

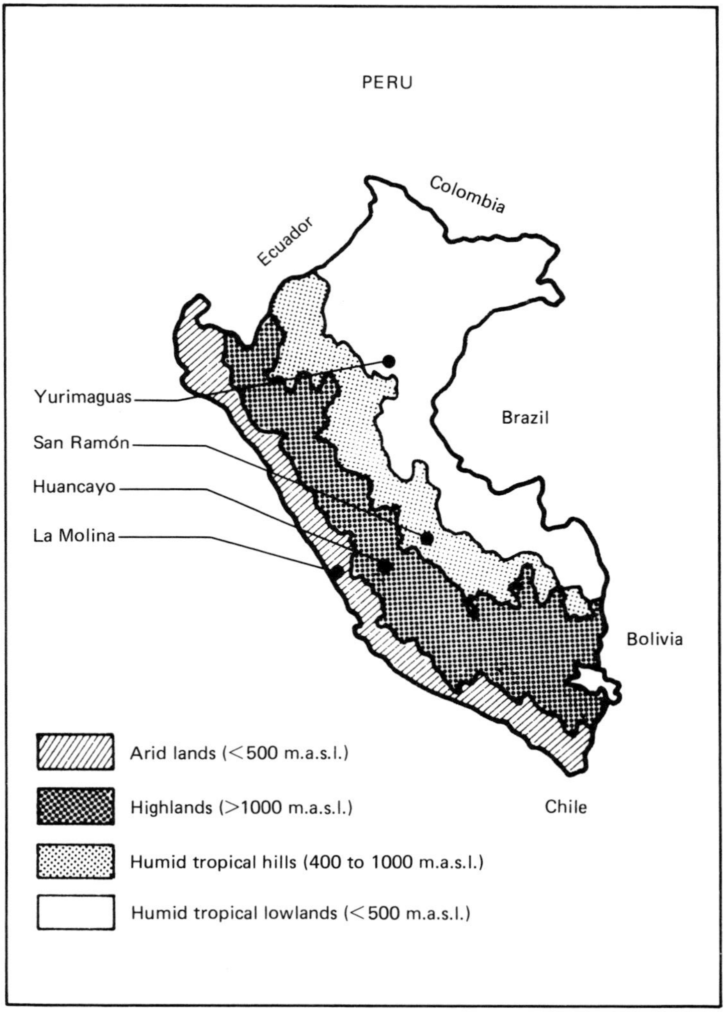

In Peru's lowland humid zone, the Amazon basin region, potatoes are imported over great distances at high cost from highland production zones, or even from wholesale markets in Lima. Their high cost makes them a luxury vegetable, available only to wealthy urban residents, but relative unfamiliarity with the potato in addition to abundant availability of other local root crops causes potato consumption generally to be quite low, often as little as 1 kg/person per year.

In the larger cities of Peru's dry coastal zone, potatoes play a regular role in the diet, due to their fairly constant market availability at moderate prices. Potatoes are co-staples with foods such as noodles, rice, sweet potatoes and bread. Consumption levels are between 50 and 100 kg/person per year. For rural and small-town coastal residents, potato consumption is often highly seasonal, occurring only immediately after the cool season harvest. During the hot season potatoes may be either unavailable or sold at high prices.

How and why potatoes are consumed

The case study summaries above indicate the quantities of potatoes consumed and the consumer groups. The conclusion is that there is great diversity in the manner in which people consume potatoes, but if attention is shifted to the features common to all the potato consumers, they can be grouped according to 'how' potatoes are consumed. 'How' does not refer specifically to the means of preparation, but to the quantity consumed at any one meal, the frequency of consumption and the relationship of potatoes to other food items comprising the meal. Figure 6.7 depicts the four positions potatoes can occupy in a meal. Each large circle represents the plate or bowl holding the principal or staple food item(s) for the meal. Smaller bowls represent major or minor side dishes that complement the staple food items. An 'X' marks the place of potatoes in each hypothetical meal. Potatoes may occupy the entire plate (Figure 6.7(a)), if they are the main staple in the diet. This pattern was common in Ireland prior to 1845, when an average of 3.5 kg and up to 7 kg of potatoes per day were consumed by each Irish peasant, who had little else to eat (Woodham-Smith, 1962). Today, the highland peoples of Rwanda, Nepal, Tibet, China, Peru, Bolivia and Ecuador also exhibit the same pattern, adding a variety of spices and sauces to the ubiquitous plate of potatoes.

The potato and other co-staples may be distributed equally (Figure 6.7(b)). This pattern is common in European countries and in the United States, as well as in segments of developing country populations.

An example of this pattern is often found in Peru's city markets, where potatoes, rice and wheat noodles may all be served on one plate, in equal portions, for a fixed price. For urban people in most of South America, potatoes, rice, noodles, cassava, sweet potato and plantains are all considered optimal counterpart staples to be consumed with meat, fish, poultry, eggs or legumes.

The potato may serve as a major side dish to the principal staple(s) (Figure 6.7(c)). In this case, potatoes complement the meal, but do not replace the staples. This pattern is seen in Central America where corn, tortillas, and beans form the mainstay of the diet, and potatoes are a 'vegetable' side dish, consumed on a regular basis, but in comparatively smaller quantities and not every day. The same pattern is reported for South Asia, in India or Bangladesh, where potatoes are one of the major vegetable complements to the staples of rice, wheat and legumes.

Finally, Figure 6.7(d) represents the instances where potatoes are only one small item among many other dishes prepared to complement the basic staple element(s). This pattern is common to Indonesia, Malaysia, the Philippines, China and other Asian and South-east Asian countries. The quantity that any one person consumes may be quite small, but the frequency with which they appear in the meal may be fairly high.

Figure 6.7. Four possible positions of the potato on a table during a meal. For explanation, see the text.

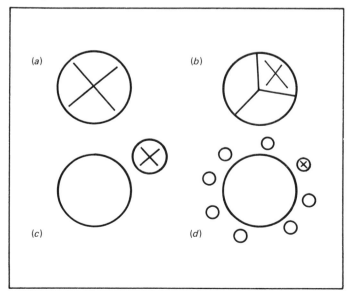

A typology of potato consumption

Observed patterns of consumption may be used to construct a typology of potato consumption, as in Table 6.5. Role 1 combines patterns (*a*) and (*b*) from Figure 6.7; role 2 combines part (*c*) and (*d*). Role 3 is appropriate when potatoes are consumed as in Figure 6.7(*c*) or (*d*), but in greatly reduced quantities and frequencies (e.g. on special occasions or as an infrequent luxury). Finally, role 4 applies when no potatoes are consumed. Cross-culturally, there are certain views, concerning potatoes, that are linked to each role. When potato is consumed as a staple it is usually regarded chiefly as an energy source. It may be grouped with grains as well as other tubers, and thought of as 'over-sized grains' or placed in a special category, but it is never grouped with vegetables such as greens, cabbage, lettuce, or radishes. The statement 'a meal without potatoes is not a meal' is commonly heard in association with role 1.

In contrast, potatoes considered to be complementary vegetables or special foods are not judged as having 'food or energy value'. In many Asian countries, it is often said that 'rice is food' and all other items, including potatoes, are merely 'trimmings, garnishings, or flavourings to rice'.

In the Philippines: 'If a man eats rice for breakfast, he can work all day, but if he eats potatoes, he is hungry by 10 o'clock.' In this role, potatoes

Table 6.5. *Typology of potato consumption*

Role	Frequency	Rate	Beliefs
1. Staple or co-staple food	5+ meals/week	±60–200 kg/year	Potato = food or energy. Meal is incomplete without potatoes
2. Complementary vegetable	1–7 meals/week	±10–60	Potato ≠ food or energy. Potato = vegetable. Potatoes are harmful if consumed frequently
3. Luxury or special food	1–12 meals/year	±0.1–10 kg/year	Potato ≠ food or energy. Potato = special food. Potato = rich people's food
4. Non-food	Never consumed	—	Potato = food for other people. Potato is unknown

are classed with vegetables. On the island of Luzon, in the Philippines, they are classed with other highland, cool-season, 'high status' vegetables such as cabbage, cauliflower, carrots, cucumbers, and celery. In the Baguio city market, potatoes may be pre-washed or plastic wrapped, as evidence of their high status. In the southern islands of the Philippines, potatoes are not so carefully displayed, and are often found among onions, garlic, ginger, and sometimes other root crops.

Associated with role 2 are certain beliefs about harmful effects of potatoes. In Guatemala these effects are part of the 'hot–cold' food philosophy, in which a cold state is attributed to potatoes since they come from the ground. This does not refer to temperature but to intrinsic qualities of specific foods. Too much of a 'cold' food or too frequent ingestion of such items is considered to harm or shock a warm body. To be safe, they must be eaten in smaller quantities than, and combined or counter-balanced with, 'hot' foods such as maize or beans.

Among such groups of people, potatoes are special foods (role 3), due either to their high cost or to certain social significance attached to them, sometimes related to the foreign cultures who first introduced potatoes. In the Philippines, potatoes are served at parties, special events such as birthdays or weddings, and at Christmas or Easter feasts. In Indonesia, potatoes are prepared in lavish dishes and exchanged to repay or cement social ties during the Lebaran celebrations which follow Ramadan, the Moslem month of fasting.

In the final category, role 4, complete unfamiliarity with potatoes is restricted to certain isolated groups and lowland areas, where lack of availability or extreme cost prohibit their consumption among all except the wealthy. There are, of course, individuals among consuming populations who, due to medical, dietary or idiosyncratic reasons, do not eat potatoes, but they are relatively few and far between. Such people usually admit, however, that potatoes are 'food' – at least for others.

Consumption role and potato prices

The above typology can be used to predict how consumer groups will react to changes in price of potatoes. Figure 6.8 shows the proposed responses for the first three role groups of Table 6.5. Role 1 consumers eat over 200 kg each per year when potatoes are freely available and cheap. Irrespective of cost, they do not consume less than 60 kg/person per year because potatoes are considered to be a basic necessity.

When potatoes are consumed as a complementary vegetable, consumption is highly responsive to price fluctuations but maximum and minimum limits of 60 kg and 10 kg/person per year are proposed. If potatoes are cheap and available, they will be consumed more frequently, if expensive less frequently, but the intake at a given meal remains fairly stable. Consumers in this role with average annual intakes in the upper ranges (35 to 50 kg/person per year) commonly state that they do not want or need to consume any more potatoes even if the price is very low.

The most interesting pattern is that of the luxury vegetable category. People buy large quantities of potatoes at special times to prepare special foods. Vendors know that potatoes are status symbols and must be obtained, so they increase their prices to their highest annual levels because they are confident that they will be able to sell all their stock. To meet this increased periodic demand, farmers delay harvesting and merchants hoard potatoes to bring prices up. After these times, prices drop again but so does demand. This results in the 'blip pattern' shown on the figure.

In summary, Figure 6.8 shows that consumption roles are not created by price levels, but rather they respond to changes in price, according to the nature of each role.

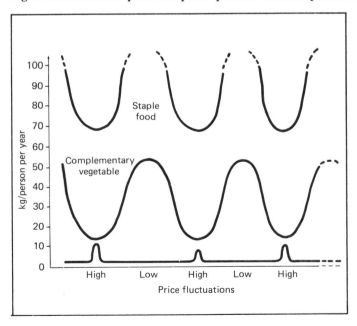

Figure 6.8. Relationship between potato prices and consumption role.

Other factors influencing consumption 211

Other factors influencing potato consumption roles

Figure 6.9 shows factors which can influence the quantity and frequency of potato consumption. Price is placed nearest to the central box because it is perhaps the most obvious factor, the easiest to measure and often thought to be the most important. In fact, as established in the previous section, price is secondary to role pattern. The other factors arranged around the outer circle act in groups to formulate the specific role. In addition to those discussed above, the following circumstances also have effects:

Historical events

When introduced in Rwanda, potatoes were considered to be taboo and were not consumed. If a person ate them it was believed that their cows would become sick and die, or that the milk would go bad. Tribal leaders, seeing that potatoes caused no harm to foreigners, especially missionaries, and had good production potential, forced local headmen to consume them and thus proved that no harm occurred to people or cattle. Forced labour migrations of people from consuming communities promoted the spread of potato cultivation and consumption.

Figure 6.9. Summary of the contributory factors determining the pattern of consumption of the potato in the human diet.

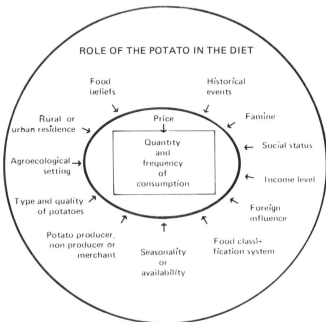

Famine

In the highlands of both Rwanda and Nepal, there was rapid acceptance of potato as a staple because little else could grow so well at higher altitudes and there had been previous periods of famine.

Social status

Potatoes in South-east Asia are expensive relative to other staple foods or native vegetables; it is assumed that if you eat them you must be wealthy. They are also symbolic of western expatriates, who normally enjoy high social status while residing abroad. Conspicuous consumption of potatoes at special occasions is therefore a way of demonstrating to others that a high level of social status has been achieved. In Singapore the younger generation is reported to be changing from eating only rice to eating potatoes as well: potato consumption is viewed as symbolic of modern ideas.

Foreign influence

Potato-consuming colonizers of developing countries have greatly influenced local consumption habits. Colonial administrators often introduced the potato and its cultivation where it was previously unknown. In Rwanda, Belgian school-masters and missionaries encouraged consumption among boarding-school children. The Dutch left behind, in Indonesia, varieties such as 'Eiggenheimer' which are still widely grown, as well as a preference for white-skinned, yellow-fleshed potatoes. The English influenced consumption in Kenya as well as throughout South Asia. Potato production and consumption increased in the Philippines only after the American occupation in 1898, although the Spanish had introduced potatoes over a century earlier.

Type and quality of potatoes

The types and qualities of potatoes can influence the consumption role. Conversely, the role can respond to variations in type or quality, according to a range of acceptable types, with rejection of those that fall outside this range. Among Filipinos, marked preferences exist for red-skinned or white-skinned potatoes, depending on the part of the country. In Luzon, red-skinned tubers are used for cold salads, while white-skinned ones are for cooking with meat or vegetables. In Mindanao, white-skinned potatoes are eaten, and red ones are often avoided because they are said to spoil quickly. In Indonesia, red- or purple-skinned potatoes are produced in a few isolated areas, but they have little market value because consumers think they look like sweet potatoes, a low-status food, and they will not buy them. For years in Guatemala, Santa Rosa was a leading potato producing area. Its potatoes were referred to as '*papa* Santa Rosa' and were highly esteemed by urban consumers. Today Santa Rosa no

Other factors influencing consumption

longer produces many potatoes, but merchants, desiring better prices for their potatoes, sell them as Santa Rosa potatoes no matter what their variety or production location, and consumers pay high prices for them. Peruvian consumers are prepared to pay more for native varieties than for improved commercial varieties (Figure 6.10).

Figure 6.10. Potato preferences. Consumers in Lima, Peru, will pay higher prices for deep-eyed native potatoes from the Andean highlands than for improved commercial varieties grown on the coast.

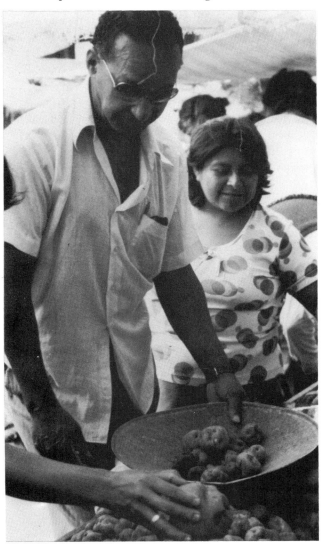

Patterns of potato consumption

The degree of importance attached to tuber size, shape, skin or flesh colour or cooking quality depends greatly on the specific consumption role and the particular consumer group. For some, small potatoes are preferable to large (Figure 6.11). Others desire a potato that holds a cube-shape when cooked, while some like potatoes that fall apart when cooked. Skin quality may be important to the staple consumers who eat the entire potato. Above all, in each group, consumers search for potatoes with the best flavour, only then identifying these potatoes with names, origins, sizes, shapes or distinctive colours. Any new potato variety must not only produce well, but fit within the local ranges of acceptable taste.

Prospects
Potential for changes in consumption roles

If all the influencing factors are taken into account, there is potential for luxury potato consumers to become complementary vegetable consumers but it is less likely that the latter will begin to utilize the potato as a staple food. As a luxury vegetable, the potato has a high status position in the diet, and it might be included in ordinary meals, if it were

Figure 6.11. Potato preferences. Consumers in central Java, Indonesia, prefer small and medium-sized potatoes rather than large ones.

more readily available or cheaper. It is much more difficult to change potatoes from a complementary vegetable position to that of staple. Usually, in the vegetable role, other foods such as grains, legumes, or other tubers are considered to be staples and are more readily available. This does not mean that the change cannot occur (and it does), but when it happens, the potato fills a gap that was either previously unoccupied or poorly occupied by local staple foodstuffs.

A goal for the future should be to facilitate greater *frequency* of consumption among those who regard potatoes as vegetables by working:

- (i) to reduce the price of potatoes relative to other vegetables;
- (ii) to encourage wider market distribution on a more regular basis;
- (iii) to improve home potato storage;
- (iv) to promote varieties suitable for local vegetable preparations;
- (v) to provide information on new methods of incorporating potatoes into local diets.

In line with the last of these recommendations, processing and weaning-food preparations are two methods to encourage greater potato utilization.

Processing

The nutritional details of the various methods of processing were considered in Chapter 4.

If consumption trends for processed potatoes among higher-income groups in the developing countries are similar to those in developed countries, it can be assumed that demand for such items will increase. However, potato processing has a much broader applicability. It can provide an alternative to storing fresh potatoes. Transforming them into flour, starch, dried or freeze-dried products makes them easier to store for long periods and less expensive to transport, and often increases their unit value. Processing also offers a means to utilize poor-quality potatoes, and to market them at a higher price than if they are sold as animal feed.

There is a pressing need for simple processing plants which can be run inexpensively, with small capital investment, at the village or commune level, by relatively inexperienced operators. Such plants have been developed in Peru and Bolivia (Figures 6.12 and 6.13). In Peru, traditionally processed dried potatoes form a significant part of the highland diet. Similar products might have a place in the diets of non-Andean populations.

Not all items are easily, or justifiably, transferable from Western culture. In Guatemala, processed potatoes fit well into several dietary niches. French-fried potatoes are rapidly becoming the preferred complement to local fried chicken, *pollo campero*. Potato chips are sold in urban areas on street corners, and at cinemas, and as a part of the general 'street food' consumed by the urban workforce. One of the more promising items currently being developed locally is instant mashed potato. *Puré de papa*, or homemade mashed potato, is already one of the most popular urban potato preparations. So a new, easier preparation, if moderately priced, has a good chance of becoming a success. Promotion of other items, such as potato flour to supplement wheat or corn flour, could meet with greater resistance, since they are not customary practices and would alter the nature of the traditional staple food items.

In Indonesia it is difficult to find any examples of processed potatoes. However, many other food items, such as cassava, rice, nuts and plantains are processed into a variety of crunchy *krupuk*, a kind of giant spicy chip, eaten with meals. Many other kinds of fried snack food are sold and consumed among all socio-economic groups. Occasionally, potatoes also are processed into chips. Possibly potatoes could play a more significant role in the *krupuk* industry. However, at present, few potatoes are available for processing, since virtually all of the fresh commodity can be sold profitably in local markets. It is also possible that the market is

Figure 6.12. Low cost, simple machinery facilitates peeling (left) and slicing (right) of potatoes in the International Potato Center's pilot processing plant at Huancayo, Peru. Intermediate technology such as this provides a viable processing alternative in areas where large-scale industrial processing is not feasible.

Prospects 217

already saturated with processed food items and there is no need for processed potatoes.

Processing holds much potential for diversifying potato consumption in the future. Caution must be exercised when introducing processing technologies in that: (1) there should first be a place for the resulting product in the diet or the possibility of developing that place, and (2) the technology used must be appropriate to the overall farming and post-harvest system.

Weaning foods

One reason for malnutrition in developing countries is a lack of good-quality weaning foods (Mudambi & Rajagopal, 1980). Potatoes, either in a fresh state or after processing, mixed with other foods or made denser calorically through the addition of oils, provide an excellent weaning food for infants. Among many mothers who are potato producers or merchants in Central America and South-east Asia, potatoes are already used as a weaning food. However, where the crop is a relatively recent introduction, potatoes have only a small role in the weaning diet.

Figure 6.13. Low cost, simple improvements enhance solar drying of potatoes at the International Potato Center's pilot processing plant at Huancayo, Peru.

Figure 6.14. Women in Bhutan caring for the potato plants on land surrounding their homes (top) and women marketing potatoes in Indonesia (bottom) are among those who can be encouraged to use potatoes in their children's weaning food.

Among many Filipino producer families, potatoes are only rarely given to children. Instead, costly items are purchased for baby food, rice and sweet potatoes are imported from the lowlands, and potatoes are overlooked. Due to the bland flavour, moderate fibre content, facility of preparation, protein quality, and balanced ratio of protein to carbohydrate energy, potatoes should be encouraged as an infant weaning food (see Chapter 3).

In future, efforts should be focused on newcomers to potato production to encourage greater use of their product in their own diets and particularly those of their children. This will require innovative educational procedures and village-level pilot projects since food habits are the cultural traits most resistant to change. Yet, if potatoes are to be more widely consumed among these regions of the developing world, then efforts must begin with children, and particularly children of producers and marketing people for whom price and availability are not constraints to increased consumption (Figure 6.14).

The place of potatoes in vegetable production

It must be emphasized that for a large part of the developing world, the potato is considered to be a vegetable. This classification often makes the potato seem to be much less important than cereal foods or other roots and tubers consumed as staples. Within such a system, it is difficult for governments to justify specific programmes dealing only with potatoes, and even more difficult to allocate personnel to work solely on potatoes. Instead, potato research and extension must be combined with that of other vegetable crops such as tomatoes and cabbages. Particularly in South-east Asia, potato improvement should be promoted within a vegetable farming system. We must be concerned not only with improving the quality of potatoes but also with improving potato production relatively, within highly intensive vegetable-cropping systems.

Conclusions

Other chapters in this book have detailed the advantages of potatoes as a cheap, nutritious, easily grown food supply. The ultimate goal of improving potato production in the tropics is to facilitate greater and wider incorporation of potatoes into existing dietary patterns. However, potatoes should not be considered in isolation. They must be viewed as a part of the larger food network, an addition, a complement, but not a substitute for, other equally beneficial foods. Viewing the potato as an exclusive monocrop or 'mono-food' will not encourage its viability in the long run. Single-mindedness can cause disaster just as single-minded dependence on potatoes reduced the Irish population by half in the nineteenth century. However, when some potato is regularly found in the daily bowl of food where it never was before, a significant goal will have been achieved.

References

Berg, A. (1981). *Malnourished people: a policy view.* The World Bank, Washington, DC.

Chavez, M. & Huenemann, R. (1981). Evaluation of household dietaries. Paper presented at the United Nations University Workshop on Methods in Large Scale, Nutritional Intervention Programs, September.

Christakis, G. (ed.) (1978). *Nutritional assessment in health programs.* American Public Health Association, Washington, DC.

Christiansen, J. A. (1980). [*The potato: production of improved seed.*] In Spanish. ICTA/PRECODEPA, Guatemala.

Den Hartog, A. P. & van Staveren, A. (1979). *Field guide of food habits and food consumption*, International Course in Food Science and Nutrition Papers, no. 1. Wageningen.

Farnsworth, H. C. (1961). Defects, uses and abuses of national food supply and consumption data. *Food Res. Inst. Studies* **2**: 179–201.

Indonesia (1978). [Food Balance Sheet in Indonesia.] In Indonesian. Biro Pusat Statistik, Djakarta.

INCAP/ICNND (Institute of Nutrition of Central America and Panama and the Interdepartmental Committee on Nutrition for National Development) (1971). *Nutritional evaluation of the population of Central America and Panama, 1965–67. Regional Summary*, USAID-DHEW Publication no. (HSM) 72-8120.

International Potato Center (CIP) (1982). *World potato facts.* Social Science Department, Lima.

Mudambi, R. & Rajagopal, M. V. (1980). Role of potato in child nutrition, *Indian Food Packer*, September–October: 25–32.

Potts, M. J. (1981). *Farm-level research to optimize potato productivity in the Philippines: some reflections.* Social Science Planning Conference, International Potato Center, Lima.

Rhoades, R. (1982). The art of the informal agricultural survey. Training Document 1982-2. Social Science Department, International Potato Center, Lima.

Ruano, S. & Calderón, S. (1981). [Basic interview techniques in carrying out farming systems research.] In Spanish. Paper presented in the Dominican Republic, March 27–30.

Scott, G. J. (1981). Potato production and marketing in Central Peru. Ph.D. thesis, University of Wisconsin, Madison.

Susenas, V. (1980). [*Organization of household catering expenditure.*] In Indonesian. Biro Pustat Statistik, Djakarta.

Werge, R. (1979). Potato processing in the central highlands of Peru. *Ecol. Food Nutr.* **7**: 229–34.

Woodham-Smith, C. (1962). *The great hunger: Ireland 1845–1849.* Harper & Row, New York.

Index

adolescents
  percentage provision of RDA for, 45
aluminium, 16, 47
amino acid content
  analyses and scores, 66–8, 69
  canning losses, 137
  and cooking, 106, 111, 112
  genetic determination, 61
  nitrogen fertilization of soil, 64–6
  potato protein concentrates, 76
  storage, 92–3
  true protein, 63
amino acids, essential (EAA), 61, 67
  indices, xv, 61, 67, 111
  non-protein nitrogen, 62, 63
  potato protein concentrate, 77
  potatoes, compared to other plant foods, 30
amino acids, free
  browning reactions, 133
  chips production, 129
  and drum-drying, 132
  french fries production, 124–5
  non-protein nitrogen, 62, 63–4
amino acids, limiting, 67
amino acids, sulphur
  and enzymic browning, 106–7
  see also methionine
amylase, pancreatic, 104
β-amylase, 10
amylopectin, 10
amylose, 10
anthocyanin, 7, 14
ascorbic acid
  and iron availability, 48–9
  forms of, 15–16
  stability, 108
ascorbic acid content
  boiled tuber and RDA provision, 45
  canning, 37, 137–8, 141

chips, 130, 141, 142
chuño, 37, 149, 150
compared to other crops and plant foods, 25, 27, 42, 44
dehydrated products, 133–4, 135, 141, 142
dietary contribution, 45–6
domestic preparations, 112–13
and freezing process, 37, 127–8
and french fries production, 125–6
papa seca, 37, 149, 150
peeling, 102, 120–1
processed products and RDA provision, 143
soil fertilization, 42–3
storage, 93–6, 99, 135
and sulphiting, 121
variation, factors affecting, 41–3
ascorbic acid content, and cooking
  dried products, 135–6
  methods of cooking, 35, 112–14, 139, 142
  peeled *v.* unpeeled, 107–8, 109
  potatoes and other plant foods, 29
ash content, 16

Bangladesh, 173, 196, 207
bean (*Phaseolus vulgaris*)
  composition, 26–7, 28–9
  dietary fibre, 39
  essential amino acids, 30
  lectins, 185
beliefs about food, 208, 209, 211
Bhutan, 193, 218
biological value (BV)
  chuño, 151–2
  definition, xv
  papa seca, 151–2
  potato protein, 70, 72, 73
  processing-waste proteins, 76

Index

biotin, 15, 43
bitter potatoes, 144, 177
   glycoalkaloids, 164, 165, 171, 177
blackening, post-cooking, 14
blanching
   and nitrogenous constituents, 124, 131–2
   and processing, 125, 132, 133
   and vitamin content, 125, 126
boiling, peeled *v.* unpeeled, 103–10
   ascorbic acid, 107–8
   carbohydrate, 104–5
   dietary fibre, 107
   minerals, 110
   moisture, 104
   nitrogenous constituents, 106–7
   vitamins, 108–9, 139
   see also peeling
Bolivia, 177, 206
boron, 16, 47
bread
   composition of, 28–9
   energy supply, 25, 28
   fibre content, 28, 39, 40
   and potato flour, 131
   and potato protein concentrates, 77–8
breast milk
   net dietary protein calories percentage, 31, 32
breeding programmes and glycoalkaloids, 164–5, 169, 186
   disease and pest resistance, 170, 186
browning, *see* discoloration
BV, *see* biological value

calcium content, 16, 47, 50
   compared to other crops and plant foods, 25, 27, 43
   and cooking, 29, 35, 110, 116–17
   and preparation methods, 142
   processing methods, 37, 102, 141
   see also mineral content; minerals
calories, *see* energy values
canning, 136 ff
   minerals, 138–9, 141
   nutrient composition, 140–1
   processes involved, 136, 137
   protein content, 137, 140
   and RDA provision, 143
   vitamin content, 137–8, 139, 141
carapulca, 149
carbohydrate content, 23
   chuño, 148, 150
   compared to other crops and plant foods, 24, 26
   and cooking methods, 28, 34
   non-starch polysaccharides, 10–11
   papa seca, 148, 150

   processing methods, 36, 140
   and storage, 90
   sugars, 11–12
   see also starch
carboxypeptidase inhibitor, 181, 182
   heat stability, 183
β-carotene content, 14, 15
   compared to other crops and plant foods, 25, 27, 29, 42
   and cooking, 29
carotenoids, 14
cassava
   composition, 24–5, 28–9
   net dietary protein calories percentage, 32
cassava flour
   protein supplements for, 33
cell walls, 8, 10–11
cellulose, 11, 39, 107
α-chaconine, 163
   and cooking, 165, 169
   in potato species, 167–8
   toxicity, 179
β-chaconine, 168
children
   education, and potato usage, 220
   percentage provision of RDA for, 45
   potato as protein and energy source, 25–6, 31–3
   potato protein nutritive value, 72–4
   protein-rich staple supplement combinations, 31–2, 33
   starch digestibility, and cooking, 104–5
China, 85, 206, 207
chips (crisps), 5–6, 128–31
   ascorbic acid losses, 130, 139
   composition, 34–5, 140–1, 142
   from unpeeled potatoes, 130–1
   glycoalkaloid content, 176
   minerals, 130, 141, 142
   nitrogenous compounds, 129
   production process, 128, 129
   and RDA provision, 143
   sugars, and colour and flavour production, 11–12
   value as snack food, 130
   vitamins, 130, 139, 141, 142
chips, *see* french fries
chlorogenic acid
   and tuber discoloration, 14, 106
chlorophyll, 15
   greening, 172
   production, and irradiation, 172
cholinesterase inhibition, 178, 179
chromium, 47, 50–1, 110
chuñificación, 144
chuño, 118, 143–4
   preparation and uses, 144–8
chuño blanco, 145

chuño blanco—continued
  composition, 35, 36–7, 148–9, 150
  glycoalkaloids, 144, 177
  nitrogen loss, 149, 151–2
  production method, 145–8, 177
chuño negro, 144, 145
  composition, 36–7, 148–9, 150
  glycoalkaloids, 144, 177
  nitrogen loss, 149, 151–2
  preparation and uses, 148, 177
chymotrypsin inhibitor, 181, 182
  heat stability, 183
  and nitrogen utilization, 183, 184
citric acid, 9, 14
cobalt, 16, 47, 51
cocoyam composition, 24–5
  see also taro
commersonine, 163, 165
consumer groups
  ethnic groups, 198
  expatriates, 198, 200
  hotels and restaurants, 198–9
  non-potato farmers, 197
  nutritional and agricultural professionals, 199
  potato marketers, 197
  potato producers, 195, 200–2
  pre-school children, 198
  urban residents, 197, 201, 204, 205, 206, 207
consumer surveys
  Guatemala, 203–4
  Indonesia, 199–200
  Peru, 204–6
  Rwanda, 202–3
consumption patterns, 191 ff
  consumer groups, 195–9
  consumer surveys, 199 ff
  food balance sheets, 192–4, 194–5
  and mealtime presentation, 3, 4, 206–7
  nutrition surveys, 194–5
consumption roles, 208 ff
  and beliefs, 208, 209, 211
  and famine, 212
  foreign influences, 212
  potential for changes, 214–15
  prices of potatoes, 209–11
  and social status, 212
  type and quality of potatoes, 212–14
cooking processes
  ascorbic acid content, 35, 112–14, 139, 142
  B-group vitamin content, 115–16
  composition, 28–9, 34–5
  dietary fibre content, 28, 34, 39, 112
  energy content, 24, 25, 28–9, 34
  and glycoalkaloids, 165, 169
  high altitudes, and proteinase inhibitors, 184
  mineral content, 35, 116–17, 142
  nitrogenous constituents, 111–12
  non-starch polysaccharides, 11
  number of stages, and nutrient content, 110–11
  peel nutrient content, 102, 103
  post-cooking blackening, 14
  protein content, 28, 34, 59, 111–12
  and proteinase inhibitors, 183, 184
  starch digestibility, 27, 104–5
  starch granules, 10
  texture characteristics, 11
  see also boiling, peeled *v*. unpeeled; micro-wave cooking; processing, large scale
copper content, 16, 43, 47, 50–1
  preparation methods, 110, 116–17, 142
corn composition, 26–7; *see also* maize
crisps, *see* chips

dehydrated products, 131–6
  composition, 36–7, 140–2
  cooking and ascorbic acid content, 135–6, 139
  methionine content, 72, 133
  mineral content, 136, 142
  nitrogenous constituents, 131–3
  production processes, 131, 132, 133
  and RDA provision, 143
  and solar energy, 118, 217
  uses, 131
  vitamin content, 133–6, 139, 141, 142
dehydroascorbic acid, 15–16, 94, 113
demissine, 163
  in potato species, 168
digestibility
  nitrogen, 64
  protein, 68, 70–1, 73
  starch, 27, 73, 104–5
diketogulonic acid, 15, 94
discoloration
  amino acid content, 106–7, 133
  chlorogenic acid, 14, 106
  enzymic, 13–14, 106–7
  inhibition by sulphiting, 121
  non-enzymic, 14
  storage temperature, 86
dormancy, 86
dry matter (DM), 8, 10
  content, 9, 23
  measurement, 9
  nutritive value, 23
  tissue distribution, 8–9
  total nitrogen correlation, 23
  variability, 8, 23

EAA, *see* amino acids, essential

Index

Ecuador, 206
energy density
  v. energy requirements, 25–7, 51
energy values
  chuño, 148, 150
  compared to other food plants, 24–6, 28
  and cooking, 24, 25, 28–9, 34–5
  papa seca, 148, 150
  processed products, 35, 36, 140
  starch digestibility, 27, 104–5
enterokinase inhibitors, 181
enzymes, 13–14
  cholinesterase inhibition, 178, 179
  lipid-degrading, 12–13
  sweetening in storage, 13
  tuber discoloration, 13–14, 106–7
  see also proteinase inhibitors

fat content
  chuño, 148
  compared to other crops and plant foods, 24, 26, 28
  and cooking, 28, 34
  dry matter, 9
  papa seca, 148
  processing methods, 36, 140
  see also lipids
fibre, crude, 38
  chuño, 148
  compared to other crops and plant foods, 24, 26, 28
  and cooking methods, 28, 34
  papa seca, 148
  in peel, 102, 103
  and processing methods, 36, 140
  and storage, 93
fibre, dietary
  compared to other crops and plant foods, 24, 26, 28, 39
  and cooking methods, 28, 34, 39, 107, 112
  definition, 38
  determination, 38
  disease protection, 38
  in dry matter, 9
  non-starch polysaccharides, 11, 38
  and peeling, 39–40
  and processing methods, 36, 140
  recommended intake, 39
  resistant starch, 38
  and storage, 93
  types of, and physiological effects, 38–9
flavours
  and glycoalkaloids, 170–2
  and lipid degradation, 12–13
  and phenols, 171
  potato protein concentrates, 77
  and sugars, 11, 12
  see also bitter potatoes

fluorine, 47, 51
folic acid
  stability, 108
folic acid content, 43
  boiled tuber, and RDA provision, 45
  canning, 37, 138, 141
  and cooking, 35, 109, 116, 139
  dehydrated products, 37, 134, 135, 141
  dietary contribution, 46
  french fries production, 126
  peeled *v.* unpeeled, 109
  processed products, and RDA provision, 141, 143
  and storage, 98
  variation, 40, 41
food balance sheets, 192–4, 194–5,
  potato consumption Indonesia, 199
  unreliability of, 192–4
french fries (chips), 5–6
  colour production, 11–12
  composition, 34–5, 142
  vitamins, 35, 139, 142
french fries, frozen, 123–7
  mineral content, 126–7, 141
  nitrogenous constituents, 124–5
  nutrient composition, 140–1
  production process, 123–4, 125
  and RDA provision, 143
  vitamins, 125–6, 139, 141
frozen products, 123–8
  ascorbic acid content, 37, 127–8
  composition, 36–7
  french fries, 123–7
  reheating, and vitamin losses, 128
  total nitrogen, 127
  vitamins, 37, 127, 139
  see also chuñificación
fructose, 11

glucose, 11
glycoalkaloids
  accumulation prevention, in tubers, 179–81
  chemical nature, 163
  chuño production, 144, 177
  and cooking, 165, 169
  disease and pest resistance, 169–70, 186
  distribution in potato plant, 163–4
  distribution in tuber tissues, 164, 165
  and environmental conditions, 174–5
  extraction and analysis, 170
  and flavour, 170–2
  and greening, 172
  and light exposure, 172, 173–4
  and phytoalexins, 170
  in potato species, 163, 164, 165, 167, 168
  in potato varieties, 164, 166
  and processing methods, 176–7
  processing waste, 77

226 Index

glycoalkaloids—*continued*
  in Solanaceae, 162
  and storage temperature, 175–6
  toxicity, 77, 164, 165, 169, 172, 177–9
  and tuber damage, 175, 177
greening, 172
Guatemala, 212–13
  beliefs, and potato role in diet, 209
  potato production and consumption, 203–4
  processed potatoes, 215–16

hollow heart, 9

India, 53, 207
Indonesia, 4, 22, 196, 198
  potato consumption patterns, 199–200
  potato role in diet, 53, 207, 209
  potato size preferences, 214
  processed potatoes, 216–17
International Potato Center, xi, xiii–xiv, 2, 3, 5
  germ plasm collection, 8, 23
  simple processing equipment, 216, 217
invertase, 13
iodine content, 116
  dietary contribution, 47, 50
Ireland, 26, 206, 221
  potato famine, 19
iron
  availability, and ascorbic acid, 48–9
iron content, 16, 47
  boiled or fresh tuber and RDA provision, 45, 49
  compared to other crops and plant foods, 25, 27, 43
  and cooking, 29, 35, 48, 110, 116–17, 142
  dietary contribution, 48–9
  processed products, and RDA provision, 143
  and processing, 141, 142
irradiation
  and ascorbic acid, 94
  chlorophyll and solanine production, 172
  lysine content, 93
  and preservation, 87, 90
  vitamin content, 98

Java
  potato consumption patterns, 199–202
  potato size preferences, 214

kallikrein inhibitors, 181
Kenya, 212
krupuk, 216

lectins, 184–5
light exposure
  glycoalkaloid production, 172, 173–4

greening, 172
solanine content, 174
lignin, 11, 38, 39, 107
lipid-degrading enzymes, 13
lipids, 12–13
  degradation, and flavour, 12–13
  tuber content, 23
  see also fat content
lipoxygenase, 13
lysine availability
  and processing, 125, 137
lysine content, 61, 67
  chips production, 129
  dehydrated products, 132
  non-protein nitrogen, 63
  and storage, 93

macaroni, 28–9
magnesium content, 16
  and cooking, 110, 116–17
  dietary contribution, 47, 50
Maillard reaction, 12, 14, 112
  amino acid losses, 124, 129
  and storage, 90
maize, 47
  cooked, composition of, 28–9
  net dietary protein calories percentage, 32
  protein supplements for, 33
  see also corn
malic acid, 14
manganese, 43, 47, 50–1
  and cooking, 110, 116–17
mazamorra, 148
medical conditions
  and dietary fibre, 38
  and lectins, 185
  and raw potato starch, 104
  and sodium intake, 50
  see also toxins
melanins, 13
melanoidins, 12
methionine content, 61, 67
  chips production, 129
  dehydrated products, 72, 133
  non-protein nitrogen, 62, 63
micro-wave cooking
  mineral content, 117
  proteinase inhibitors, 183
  vitamin content, 116
millet, 33
mineral content, 16, 47
  canning, 138–9
  chips, 130
  chuño, 149, 150
  cooking processes, 110, 116–17
  dehydrated products, 136
  dietary contribution, 48–51, 53
  french fries production, 126–7

frozen products, 37, 126–7
papa seca, 149, 150
peeled *v.* unpeeled, 110
and storage, 98–9
tubers, factors affecting, 48
minerals
  dietary contributions, 48–51
  distribution in tuber, 102
  see also individual minerals
molybdenum, 43, 50–1, 116
  content, 16, 47
moraya, 144

Nepal, 19, 206
net dietary protein calories percentage (NDpCal%), 31–2
niacin
  availability, 45
  stability, 108
niacin content
  boiled tuber and RDA provision, 45
  canning, 138, 141
  chuño, 149, 150
  compared to other crops and plant foods, 25, 27, 29, 42
  and cooking, 29, 35, 108–9, 115–16, 139, 142
  dehydrated products, 134, 135, 141, 142
  dietary contribution, 46
  french fries production, 125–6
  frozen products, 37, 125, 127
  papa seca, 149, 150
  peeled *v.* unpeeled, 108–9
  peeling methods, 120, 121
  processed products, and RDA provision, 143
  processing methods, 37, 141, 142
  and storage, 97–8
  variation, 40, 41
nickel, 16, 47, 51, 110
nitrogen, Kjeldahl, 58
nitrogen, non-protein (NPN), 60
  amides, 60, 62–4
  amino acid composition, 62–4
  and blanching, 131–2
  and cooking methods, 106, 111, 112
  distribution in tuber, 59
  and environmental factors, 64
  nitrogen digestibility, 64
  ratio with protein nitrogen, and storage, 91–3
nitrogen, soil fertilization
  amino acid content, 64–6
  ascorbic acid content, 42–3
  mineral content, 48
  true protein content, 61–2
  vitamin content, 40, 42
nitrogen, total
  and blanching, 131

canning, 137
chuño, 151–2
constituents, 60
and cooking methods, 106, 111–12
correlation with dry matter, 23
and dehydration, 131–2
distribution in tuber, 59
in dry matter, 9
factors affecting, 59
french fries production, 124
frozen products, 127
papa seca, 151
in peel, 102
and storage, 90–1
nitrogen balance
  adults, 71, 72
  children, 72–3, 74
NPN, *see* nitrogen, non-protein
nutritional values
  compared to other crops and plant foods, 24 ff, 42, 43
  dry matter, 23
  underestimation, 3, 19
nutritional values, proteins
  adults, 71–2
  animal feeding experiments, 70
  children, 72–4
  digestibility, 70–1
  microbiological assays, 68, 70
  processing waste, 76, 77

oatmeal, essential amino acids, 30
oats, 32, 33
  net dietary protein calories percentage, 32
  protein supplements for, 33
organic acids, 14; *see also individual organic acids*

Pakistan, 22
pantothenic acid content, 43
  dietary contribution, 46
papa seca, 118, 145
  composition, 36–7, 148–9, 150
  glycoalkaloids, 177
  nutrient losses, 149, 151–2
  preparation, 144, 148, 177
  uses, 149
pectins, 8, 11, 39
peel, 8
  glycoalkaloids, 164, 165, 169, 186
  nutrient content, 101–3
peeling, 100 ff
  abrasion-peeling, 120
  dietary fibre, 39–40
  lye-peeling, 120–1
  nutrient loss, 101, 120–1

peeling—*continued*
    steam-peeling, 120
    weight loss, 100–1
    see also boiling, peeled *v*. unpeeled;
        pre-peeling
PER, *see* protein efficiency ratio
Peru, 52, 53
    agroecological zones, 205
    cooking, and proteinase inhibitors, 184
    mealtime presentation, 206, 207
    potato preferences, 213
    potato production and consumption, 204–6
    processed potatoes, 215, 216, 217
    storage, 89
    see also chuño; papa seca
Philippines, 180
    consumption patterns, 207, 212
    potato role in diet, 208–9
    potato yields, 192
    weaning foods, 2, 20
phosphorus content, 16, 47
    compared to other crops and plant foods, 25, 27, 43
    and cooking, 29, 35, 110, 116–17
    dietary contribution, 47, 49–50
    and processing, 37, 141
    and starch, 10
phosphorylase, 13
phytic acid, 14, 49–50
phytoalexins, 170
pigments, 14–15
    anthocyanins, 7, 14
    chlorophyll, 15, 172
    melanin, 13
    melanoidin, 12
plantain, 207
    composition, 26–9, 47
    net dietary protein calories percentage, 32
    protein supplements for, 33
polyphenoloxidase, 13
polysaccharides, non-starch, 10–11
    dietary fibre, 11, 38
    tuber dry weight, 10–11
potassium content, 16
    and cooking, 110, 116–17
    dietary contribution, 47, 50
potato fruit water, 75, 76
    toxic factors, 77
potato protein concentrate (PPC)
    composition, 76
    in human food, 77–8
    toxic factors, 77
pre-peeling, 119–21,
    nutrient content, 120–1, 139
    and sulphiting, 121–3, 139
prices and consumption, 209–11
processed products, and RDA provision, 143

processing, large scale, 118 ff
    developing countries, 215–17
    glycoalkaloid content, 176–7
    growth of, 118–19
    nutritional changes, 84, 139–42
    pre-peeling, 119–21
    pre-processing operations, 120
    sulphiting, 121–3
    vitamin losses, 139
    see also canning; chips; cooking processes; dehydrated products; french fries; frozen products; proteins, processing waste
processing, traditional, *see chuño blanco; chuño negro; papa seca*
protein content, 23, 30–6
    canning, 137
    chuño, 36, 148, 150, 151
    compared to other crops and plant foods, 24, 26, 28
    contribution to household protein intake, 31
    and cooking, 28, 34, 59, 111–12
    and energy supply, 31–3
    french fries production, 124
    nitrogen fertilization of soil, 61–2
    papa seca, 36, 148, 150, 151–2
    processed products, and RDA provision, 143
    and processing, 36, 137, 140
    staple supplement combinations, 30, 31–3
    tubers, 9, 23
protein efficiency ratio (PER), 66, 70, 72
    definition, xv
proteinase inhibitors, 181–4
    cysteine content, 183–4
    disease and pest resistance, 182
    heat sensitivity, 183
    nutritional significance, 183–4
proteins
    digestibility, 68, 70–1
    in dry matter, 9
    intake, safe level evaluation, 68
    quality estimation, 66–70
    soluble, 60, 61–2
    see also net dietary protein calories percentage; nutritional values, proteins; potato protein concentrate
proteins, processing waste
    air classification, 77
    food industry uses, 77–8
    in human food, 77–8
    nutritional value, 76
    potato fruit water, 75, 76
    recovery of, 75–6
    single-cell protein production, 75, 76
pyridoxine
    stability, 108

Index

pyridoxine content, 43
  boiled tuber and RDA provision, 45
  dehydrated products, 134, 135, 141
  dietary contribution, 46
  canning, 138, 141
  cooking methods, 35, 109, 115–16, 139
  french fries production, 126
  peeled *v*. unpeeled, 109
  processed products, and RDA provision, 143
  and processing, 37, 141
  storage, 98
  variation, 41

recommended daily allowance (RDA), percentage provisions, 45, 143
reconditioning, 12, 86, 90, 93
  ascorbic acid levels, 96
  free amino acid content, 92
  and total nitrogen, 91
riboflavin content
  canning, 138, 141
  chuño, 149, 150
  compared to other crops and plant foods, 25, 27, 29, 42
  and cooking, 29, 35, 109, 116
  french fries production, 125
  papa seca, 149, 150
  peeled *v*. unpeeled, 109
  peeling methods, 120, 121
  and processing, 37, 141
  storage, 97
  variation, 40, 41
riboflavin stability, 108
rice
  composition, 26–7, 28–9
  consumption patterns, 207
  essential amino acids, 30
  net dietary protein calories percentage, 32
  protein content, 59
  protein supplements for, 33
Rwanda, 85, 100, 206
  agroecological zones, 202
  beliefs, and potato consumption patterns, 211
  foreign influences on consumption, 212
  potato production and consumption, 202–3
  storage, 88

selenium, 47, 50, 51, 116
senescent sweetening, 12, 86, 87
sodium content, 16, 47, 50
  and cooking, 110, 116
soil conditions
  glycoalkaloids, 174
  mineral content of tubers, 48
  vitamin content, 40

soil fertilization
  amino acid content, 64–6
  ascorbic acid content, 42–3
  mineral content, 48
  protein content, 61–2
  vitamin content, 40, 42
solamarines, 163
  in potato species, 168
solanine
  content, and light exposure, 174
  and greening, 172
α-solanine, 163
  and cooking, 165, 169
  in potato species, 167–8
  toxicity, 179
solar drying, 118, 217
sorghum
  composition, 26–7, 28–9
  net dietary protein calories percentage, 32
  protein supplements for, 33
specific gravity, and dry matter content measurement, 9
sprouting
  ascorbic acid content, 96
  storage conditions, 86
  suppression, and irradiation, 90
  thiamin content, 97
starch
  distribution in tuber, 10
  in dry matter, 9, 10
  dynamic equilibrium with sugars, 11–12
  enzyme-resistant, 38, 112
  industrial uses, 10
starch digestibility
  and cooking, 104–5
  and energy supply, 27
starch gels, 10
starch grains, 8
storage, 85 ff
  chuño, 144, 177
  developing countries, 88, 89
  dormancy maintenance, 86
  and irradiation, 87, 90
  light exposure and glycoalkaloid levels, 173
  nutrient determinations, factors affecting, 87
  objectives, 86
storage, nutritional changes, 152–3
  ascorbic acid, 93–6, 99, 135
  B-group vitamins, 96–8, 99
  carbohydrates, 90
  fibre, 93
  minerals, 98–9
  protein: non-protein nitrogen, 91–3
  total nitrogen, 90–1
storage temperatures
  ascorbic acid content, 94–6

storage temperatures—*continued*
  B-group vitamins, 97–8, 99
  glycoalkaloid content, 175–6
  low temperature sweetening, 13, 86
  protein: non-protein nitrogen, 92–3
  sprouting, 86
  sugar content, 9, 11–12, 13, 86–7
  total nitrogen, 90–1
  see also reconditioning
sucrose, 9, 11
sucrose phosphate synthetase, 13
sugars
  content, and storage temperatures, 9, 11–12, 13, 86–7
  dynamic equilibrium with starch, 11–12
  low temperature sweetening, 13, 86
  see also reconditioning
sugars, reducing
  colour of products, 11, 12
  in dry matter, 9
sulphiting, 121–3
  ascorbic acid content, 121, 139
  thiamin content, 121–3, 139
sweet potatoes
  composition, 24–5, 46–7
  net dietary protein calories percentage, 32
  protein supplements for, 33

taro
  composition, 24–5
  protein supplements for, 33
temperature
  reheating, and vitamin content, 128, 136
  see also chuñificación; french fries, frozen; frozen products; storage temperatures
thiamin
  stability, 108
thiamin content
  boiled potato and RDA provision, 45
  canning, 137, 138, 141
  chuño, 149, 150
  compared to other crops and plant foods, 25, 27, 29, 42
  and cooking, 29, 35, 108, 115, 139, 142
  dehydrated products, 134–5, 141, 142
  dietary contribution, 46
  french fries production, 125–6
  papa seca, 149, 150
  peel, 102, 103
  peeled *v.* unpeeled, 108, 109
  peeling methods, 120, 121
  processed products, and RDA provision, 143
  and processing 37, 141, 142
  storage, 96–7
  and sulphiting, 121–3
  variation, 40, 41

Tibet, 206
tomatine, 163, 168
tongosh, tokosh, 148, 177
tortillas, 207
  composition, 28–9
  energy supply, 25, 28
toxins
  glycoalkaloids, 77, 164, 165, 169, 172, 177–9
  lectins, 185
  potato protein concentrates, 77
  trypsin inhibitors, 77, 181, 182, 183
  see also proteinase inhibitors
trace elements, 16, 47
  dietary contribution of potato, 50–1
  peel, 102
  soil, and tuber ascorbic acid content, 43
  tuber, factors affecting, 48
  see also individual trace elements
trypsin inhibitor, 181, 182, 183
  in potato fruit water, 77
tubers
  amino acid composition, 64–6
  nitrogen distribution, 59
  nutrient distribution, 101–3
  pH, 14
  structure, 7–8
Tunisia, 173
tunta, 144
tyrosinase, 13

vanadium, 51
vitamin content, 15–16
  analytical techniques, 44
  and blanching, 126
  boiled tuber and RDA provisions, 44–5
  cold storage, 97, 98, 99
  compared to other vegetables, 42–3, 44
  and cooking, 107–9, 116
  dietary contribution, 44–7, 52–3
  french fries production, 125–6
  losses in cooking and processing, 139
  peel, 102–3
  processed products and RDAs, 143
  and reheating, 128, 136
  variation, factors affecting, 40, 42–3
vitamins
  distribution in tuber, 101–3
  vitamin A, 46–7
  see also individual vitamins

weaning foods, 74, 105, 198, 217–18
wheat
  composition, 26–7
  essential amino acids, 30
  net dietary protein calories percentage, 32
  protein supplements for, 33

Index

wound healing, 87
wounding
 discoloration, 13
 glycoalkaloid content, 175, 176, 177
 see also peeling

yams
 composition, 24–5
 net dietary protein calories percentage, 32
 protein supplements for, 33

Zaïre, 21
zinc content, 16, 47
 dietary contribution, 47, 50, 51
 and preparation methods, 102, 110, 116–17, 142